6.95

Statistics

D0121326

The M & E Handbook Series

Statistics

W M Harper ACMA, MBIM

Fifth Edition

Pitman Publishing
128 Long Acre, London WC2E 9AN

© Longman Group UK Ltd. 1988

First published in Great Britain 1965
Reprinted 1966 (twice), 1967, 1968, 1969
Second Edition 1971
Reprinted 1972, 1973, 1974, 1975, 1976
Third Edition 1977
Reprinted 1979, 1980, 1981
Fourth Edition 1982
Reprinted 1983, 1984, 1986
Fifth Edition 1988

British Library Cataloguing in Publication Data

Harper, W.M.
    Statistics. — 5th ed. — (M&E
handbooks series, ISSN 0265-8828)
    1. Mathematical statistics
    I. Title
    519.5   QA276

    ISBN 0 7121 1995 7

Typeset by Avocet Marketing Services, Bicester, Oxon.
Printed and bound in Great Britain.

Contents

Preface to the fifth edition

Since the first edition of this book was published in 1965, I have always maintained that one should leave well alone the engine that is running smoothly. But after 20 years even the best of engines can be improved a little by a thorough overhaul. So now, in this fifth edition, I have taken the opportunity of giving the book such an overhaul. Should readers of any of the previous editions glance through this one they may well find substantial differences. It is not quite a new book – statistics is a subject that remains very much the same from one year to the next and so there are rarely new innovations to be announced and evaluated – but a slightly deeper treatment has been adopted together with something of a reorganisation of a few of the chapters. It should, therefore, recognisably be the same animal as before and hopefully as useful to students as the previous editions apparently were.

The major changes. Though nothing may be less relevant to a new reader of a book than how the current edition differs from the previous one, I feel a few words here may be of value to anyone wondering just what the differences are. Apart from the simple rewriting of the same material in places, there are two major changes.

The first of these changes relates to small sample theory. Sensibly, professional bodies for whom statistics is a peripheral subject have realised that the complexity of small sample theory really calls for this aspect of statistics to be left alone. Not only did pressure of time allow the student to obtain no more than a superficial understanding of this difficult subject, but examination questions occasionally seemed to suggest some of the examiners were equally out of their depth. So this theory has now departed from most syllabuses and consequently from this book, too.

The second change is a sadder one for, alas, the book *Basic Mathematics* is no longer in print. This has meant that I have had to bring back into this book those basic mathematical topics previously covered in that other volume. Frankly, there is little enough space in a handbook to cover statistics itself without also having to cover its supporting mathematical theory. It can also be distracting to the reader to find a purely mathematical exposition in the middle of the development of a statistical topic. For this latter reason I have relegated two mathematical topics to an appendix and away from the main text.

The age of the calculator. In the previous edition I also relegated the mathematical procedures for computing statistical measures to an appendix. Those procedures used short-cut methods – 'short-cut' in the sense of simplifying the actual arithmetic called for. However, calculators are now so ubiquitous that it is easier to apply them to calculating by the 'long' method than by the 'short'. So I have assumed that every reader has a calculator and detailed in the text procedures that minimise the work needed when using such an aid. By yesterday's standards the intermediate figures in the computation look horrific but, of course, the only figures we really look at are the ones at the end, so this is of no significance.

Statistics is more than arithmetic. Before starting the book there is one point that should be made very clear. Since the ability to analyse data mathematically is fundamental to the statistical work in most examination questions, I have concentrated on this aspect of the subject. Equally important, however, is the more subjective and intangible ability to judge the quality of the raw material analysed and the conclusions implied by the analysis. A competent statistician does not simply apply the mathematical recipes at hand to the data before him in an unthinking and uncritical manner, but is instead always alert to the possibility of the data having some subtle feature that could render the obvious conclusion invalid. Unfortunately space does not allow me to develop this important facet of the subject but the student is warned here that even should he acquire a firm grasp of the mathematics of every technique in this book, this will no more make him a statistician than a knowledge of the rules of chess will make him a Grand Master.

Examinations. Since many lecturers take considerable pains in selecting and analysing for class study the questions at the end of these kinds of books, it is annoying for them to find that the questions change at each edition – particularly as old questions are often quite as good as, and sometimes better than, new ones. And as I completely revised the appendix on examination questions in the last edition, I feel I should leave this particular part of the book relatively untouched. Of course, there are fashions in examination questions like everything else – and now rapidly changing syllabuses add a new dimension to such fashions – and to allow for this I have replaced some questions with new ones. But otherwise this part of the book, at least, is unchanged.

Progress tests. As before, all chapters end with a progress test having suggested answers given in Appendix 4. Since these tests enable me to bring out points which, while useful, are not suitable for inclusion in the text, the student is advised to study these questions and answers even if he elects not to try and answer the questions on his own. Additionally, the progress tests now include one or more assignments. These are questions for which no answers are given (except, perhaps, as a few final figures) and which essentially call for more extensive workings than normal.

Inflation. This invariably renders all price figures in a book out of date before the book is even published. I bow to the inevitability of this and have made little attempt to avoid the problem, contenting myself with merely pointing out that the application of the techniques involved is unaffected by the antiquity of any monetary data.

Approach to rigour. Finally, let me say that it has been my experience that excessive qualification of statements or completely rigorous definitions, though appropriate for experts and essential at the higher subject levels, serve only to confuse the student whose object at this early stage is wisely limited to grasping the basic principles. Since student comprehension has been my dominant aim throughout, I have endeavoured to avoid such confusion.

Acknowledgments. I gratefully acknowledge permission to quote from the past examination papers of the following bodies:

Chartered Institute of Management Accountants (CIMA);
Chartered Association of Certified Accountants (ACA);
Institute of Chartered Accountants in England and Wales (CA).

WMH
1987

Part one
Introduction to statistics

1
Nature of statistics

Statistics concerns itself with obtaining an insight into the real world by means of the analysis of numerical relationships.

Let us be quite clear about this. Statistics is not tables of numbers, sets of techniques, lists of formulae, but is an approach to understanding the world about us. However, that world is very complex and there is no quick and easy way of gaining that understanding. Unremitting care and attention are prerequisites. The figures and graphs that you put on paper are but a trivial aspect of statistics; it's seeing what lies behind the figures and graphs that calls for real statistical skill. Always in statistics you must pause and ponder: be critical of argument and sceptical of measures. And if there are times when you feel you are sinking in a sea of confusion and uncertainty, do not be discouraged for that appears to be the fate of all who seek the real world beneath its misleading surface appearance. Conversely, if everything seems simple and crystal clear then stop and reflect, for very probably you have misunderstood the issue you are considering. Paradoxical as it may seem, the statistician has much in common with the poet, for both must be able to see the deeper reality beneath the veil of the apparent.

Now, if you just wish to pass an examination you will want less of the poetry and more of the pragmatism, and this book aims to give you just that. But as you memorise formulae, practise techniques and learn the conventional answers to the conventional questions don't forget that at the end of the day it's your ability to see through, in more senses than one, the theory and figures that is really important.

Introduction

First, a few general points.

1. Statistics and numbers. Only recently has it been realised that society need not be run on the basis of hunches or trial and error. The development of statistics has shown that many aspects of progress depend on the correct analysis of numerical information and relationships, particularly in economics and business. Increasingly, numbers have become the basis of rational decisions rather than hunches and events are proving that these decisions based on numbers give better results.

All this has led to an unprecedented demand for numbers – a sort of 'numbers explosion'. But, as has already been stressed, numbers have to be understood and correctly handled, and this is the role of statistics. It is a role that will grow rapidly in importance – in business, in government and in science. It may even be one of the factors which determine the future rate of human progress.

2. Numbers and the mind. It is a true, if unfortunate, fact that the mind of man was never designed to handle numbers. Probably the highest number a man can truly grasp is five or six (number, note – not symbol for a number such as 1,000 or 10^5, for if that were so it would be quite easy to grasp infinity since this is symbolised as ∞). Beyond this level we have to resort to counting and while we may be technically able to count to a very large number we cannot say this enables us to grasp the number.

A basic function of statistics, then, is to act as a device for converting ungraspable numbers into a clearer if not perfect vision of the reality lying behind those numbers.

3. Descriptive and inferential statistics. There is, of course, a close connection between understanding a number and using that understanding to make estimates of other unobserved numbers – just as an understanding of the series, 1, 2, 4, 8, 16, 32 enables us to deduce that the next number will be 64. This feature results in the subject of statistics being divisible into two broad categories:

(*a*) *descriptive statistics* – which deals with methods of *describing large masses of numbers* (the bulk of this book, in fact);

(b) *inferential statistics* – which deals with the method of *drawing conclusions from numbers observed* (the sampling theory and correlation sections of this book).

4. Variable. A very common term in statistics which a student should learn at once is *variable* which can be defined as the *characteristic that is being measured* in the given statistical study. In statistics, by definition, numbers are used to measure characteristics such as height, weight, time, money, the number of marriages or houses, or even blue-eyed men in France. Each of these could be the variable in any given study. (This form of measurement is so common that the student may need reminding that such things as good and evil, courtesy, artistic merit, love and loyalty cannot be measured by numbers.)

There is no restriction on the number of variables we must have in any statistical analysis, though of course there must be at least one. In this book most of our attention will be directed to single variable analyses, though two-variable problems (such as the relationship between advertising and sales) is the subject of the whole of Part four.

5. Time series. Another common term that should be learnt is *time series*, which is no more than a *series of numbers relating to different periods or points in time*, e.g. marriages year by year, sales month by month, radioactivity minute by minute, or earthquakes as and when they occur. The unique feature of time in the context of statistics is that nothing can stop it or slow it down or affect it in any way. It is a variable which is independent of all and everything.

Interpretation of statistics

One of the first things any student of statistics must learn is that figures can be interpreted wrongly very easily. Hence sayings such as 'you can prove anything with figures' and 'there are lies, damn lies and statistics'. Because of their experience many people distrust statisticians and, in fact, even consider them to be naive gentlemen who cheerfully insist on talking the most patent nonsense simply because their figures appear to point in that direction. Nothing could be further from the truth – indeed, many people will be startled, even annoyed, to find how cautiously statisticians approach what appears to be the most obvious and inevitable conclusion. If, for example, it can be proved that 40 per cent of all people who live in Smithville are admitted to Smithville hospital at some time in their lives, then it would appear that if Mr Smith lived all his life in Smithville there is a significant probability that one day he will be admitted to the hospital. Statisticians would point out this is not necessarily so – particularly if the Smithville hospital were a maternity hospital.

It is necessary, then, to learn to look critically at all conclusions derived from statistics. This section is concerned with illustrating this kind of critical thought. The reader, however, should be warned against sterile caution. If 50 per cent of the population are female, the other 50 per cent must be male – postulating a third sex is ridiculous and serious consideration of such wildly unrealistic possibilities is, in fact, a greater bar to progress than making a simple mistake of interpretation.

6. Biased sources. All too often public statements that rely on statistics are unfortunately biased. To test such a statement you need to ask yourself, 'who says this?', 'why does he say it?' and 'how does he know?' It should be appreciated, though, that to say a source is biased does not mean that the statement is a deliberate lie but only that the person concerned has understandably picked the figures which will show his case in the best light.

Examples

(1) '75 per cent of the voters voted for me. Most people in this constituency, then, must want me as their MP.' What if less than two-thirds of the people voted – or half were not even on the electoral roll?
(2) 'We interviewed 500 employees in the works' canteen. 95 per cent used our product, so it must be good.' What if it were your own canteen and you operated a special discount scheme for your employees? Or even that you had a monopoly of the product (e.g. you were an Electricity Board)?
(3) 'Nine out of ten female TV stars genuinely believe our product is the best.' Did the tenth change her mind after she had been selected?
(4) 'Nobody has ever put forward a justifiable complaint as to the way we handle matters.' Who decides what is justifiable?

(5) 'All the best-looking girls in this town buy our soap.' How do you know they do? How do you know they are the best-looking? What do they do with the soap – trade the wrappings for gifts reserved for the best-looking girls in town?

(6) 'Since Mr Jones became Chief Constable arrests have gone up 50 per cent.' Maybe, but what about convictions?

(7) 'Analysis of records shows that 75 per cent of our students double their pay within ten years of taking our course.' Analysis of other records may show that age, inflation and initiative result in 80 per cent of *all* students doubling their pay within ten years of taking *any* course.

7. Invalid arguments. A source may have no bias at all yet it is still very easy for a speaker to put forward an invalid argument. To test this one should ask oneself, 'does this follow?'

Examples

(1) 'Party X has doubled its votes since the last parliamentary election. This proves that its support is greater than ever.' If, in fact, three times as many candidates stood for election then the argument is invalid. Very probably the support has always been there – but it was not until the 'last election' that it had the opportunity of showing itself in the form of votes.

(2) 'No candidate for whom Harry has voted has ever won an election. Therefore Harry can stop a candidate winning by voting for him.' Harry is the sort of person who votes for causes which are already lost – he does not make them losing causes by voting for them.

(3) 'You're safe driving with me. I've never had an accident.' Just passed your test? Besides, many accidents occur as a result of what *someone else* does.

(4) 'More people die in bed than anywhere else. Bed, then, is the most dangerous place in the world.'

(5) 'Tea kills! Questioning persons involved in serious accidents disclosed that 95 per cent had drunk tea within the previous 12 hours.'

(6) 'Not one out of 3,000 people interviewed at random had ever seen a polar bear roaming freely in England. Polar bears, then, don't roam freely in England.' There's a high probability that the

same 3,000 people would say they had never seen the Mayor of Middlewich but that would hardly prove he did not exist. Moreover, those 3,000 people could all be natives of Peru. Note here that although the argument is invalid, the conclusion is still correct. Do not, then, confuse an invalid conclusion with a false one.

(7) 'The £ went up by 50 per cent last year. This year it has dropped by only 40 per cent. So the £ is worth more now than it was at the beginning of last year.' If the £ was worth, say, 100 at the beginning of last year it ended the year at 150. If it then dropped 40 per cent it dropped to 150−40 per cent of 150=90. So it's now worth less than at the start.

8. Alternative explanations. Sometimes a fallacy in an argument is easier to detect if you ask yourself, 'is there any other possible explanation?' Any statistical conclusions depend upon certain assumptions. These assumptions may or may not be warranted. Thus, the argument at the beginning of the section regarding the chance of Mr Smith's admission to Smithville hospital assumed that he belonged to the same class of people *as regards sex* as those who entered the hospital. In fact he belonged to the same class only as regards place of residence.

Examples

(1) 'Hospital records show that the number of people being treated for this disease has doubled in the last 20 years. Twice as many people, therefore, suffer from this disease than did 20 years ago.' Possibly more people go to hospital today when they have this disease or, possibly, 20 years ago diagnosis was less accurate and they were being treated for something else.

(2) 'Theatres are fuller in London than in Paris. What nonsense to say that the English are less theatre-conscious than the French.' There could be fewer theatres in London, or possibly the theatres outside London are considerably emptier than those outside Paris.

(3) 'More professional people buy X paper than any other paper. This shows X's ideas and policies are in tune with modern professional thoughts.' On the other hand it could be that X carries the best 'Situations Vacant' column.

(4) 'More and more families travel today by car. This proves our

standard of living is going up in leaps and bounds.' Or could it be railway fares?

(5) 'Last year 700 employees produced 150,000 ladders. This year 650 employees produced 160,000 ladders. This shows we have increased our productivity.' Or decreased ladder sizes.

(6) 'Scandal! 20 per cent of the women on this estate had babies last year, but only 10 per cent of the men were married.' Could be that on the estate there are 20 married couples (each of whom had a baby), 80 single women and 180 single men.

(7) 'Mortgage conditions are causing large numbers of people to be homeless. Only 50 per cent of the mortgage applications received last year were accepted by building societies.' Possibly applicants applied twice.

(8) 'Bosses' pay went up 5 per cent last year while workers' pay went up 10 per cent. The pay gap is narrowing.' If the bosses' pay rose from £100,000 a year to £105,000 while the workers' pay rose from £5,000 to £5,500 the gap has *grown* by £4,500.

Progress test 1

(*Answers in Appendix 4*)

1. Consider the following statements critically:

(*a*) 'Five years ago the average stay of patients in this hospital was 21 days – now it is only 16 days. This shows that we now cure our patients more quickly.'

(*b*) 'Most car accidents occur within 5 miles of the driver's home. Therefore long journeys are safer.'

(*c*) '10 per cent of the drivers involved in 100 car accidents had previously taken X. A parallel survey of drivers *not* involved in accidents shows that only 1 per cent had taken X. This shows that taking X is a contributory cause of accidents.'

(*d*) 'Our existence is justified. 34 people who would have turned to a life of crime have been saved by us.'

(*e*) '1,000 boy victims of broken homes who later could be traced were rated on a "good citizenship" scale. 700 of them were above average. Broken homes clearly help make better citizens.'

(*f*) 'More people were hurt in the home than in factories and mines. Homes, then, are more dangerous places than factories and mines.'

(g) 'Of 1,000 people on this estate 75 per cent are over 21 and 50 per cent are Irish. Therefore at least 250 of these people are Irish and over 21.'

Assignment

Find in current press publications half a dozen examples of debatable statistical reasoning and conclusions.

2
Accuracy and approximation

In practical statistical work complete accuracy is rarely possible. It is necessary, therefore, to use approximations, and the use of approximations leads in turn to rounding. Although the use of approximations and rounding is common enough in real-life calculations, the implications of using these well-known techniques may not always be appreciated. Indeed, their very simplicity may well disguise the potential errors that accompany their use and, since statistical conclusions can never be better than the original data from which they are drawn, it is worth looking briefly at this aspect before moving on to the arrangement and interpretation of statistical data.

1. Statistical error. Note that the word 'error' can have two meanings:

(a) *An error can be a mistake.* So if someone says that $2+2=5$ then he has made an error.

(b) *An error can be a divergence from accuracy.* If I am measuring the length of a metal bar then no matter how careful I am the figure I record will almost certainly be not quite the exact length. This may be due to the ruler not being wholly accurate, or because I have to judge a fraction of a division on the ruler, or even because the bar has expanded slightly owing to the temperature being above that laid down in the measurement standard. Whatever the reason, there is a divergence from exact accuracy, i.e. my measurement has some degree of error.

In statistics an error arising as a result of a divergence from accuracy is called a *statistical error*, and the word 'error' in this book almost invariably refers to a statistical error.

2. Causes of error. There are two basic causes of error in real-life statistics.

(*a*) *Inaccurate figures.* In practice it is rare that completely accurate figures can be recorded. If, for example, articles are weighed there is a limit to the accuracy of the scales. And when a count is made there may well be an element of doubt about the final figure, e.g. however carefully a census is carried out there are always some people omitted for one reason or another who should have been included, and vice versa, and so the total is not completely accurate.

(*b*) *Incomplete data.* Complete accuracy is also impossible where calculations are made from data lacking all the necessary information. For example, in a later chapter we shall discuss the calculation of an average from data where individual values are not given, only the number of items which fall within a given range. Such an average figure must inevitably be only an approximation – albeit a very close approximation – to the true figure.

3. Rounding. If *some* figures in a survey will prove not to be accurate, there is little point in recording the other figures with complete accuracy. Consequently, when such figures are collected they are frequently *rounded*. For instance, in a survey of petrol sold at petrol stations the recorded figures may be rounded to the complete litre, to tens of litres, or even to hundreds of litres. This means making a decision as to *how* to round the actual detailed figures.

There are three methods of rounding.

(*a*) *Round up.* If figures are to be rounded by raising them to, say, the next ten litres, an actual figure of 185 would be recorded as 190.

(*b*) *Round down.* Conversely, the figure can be rounded by reducing it to the previous ten. In this case 185 would be recorded as 180.

(*c*) *Round to the nearest unit.* The figure can be rounded to the nearest ten. If a figure is exactly half way between, such as 185, the rule is 'round so that the rounded digit is an even figure'. Rounding 185 to the nearest ten means we must choose between 18 tens (180) or 19 tens (190). Since the rule says choose the even, 18 tens, i.e. 180, is recorded.

4. Biased and unbiased error. Although the inevitability of error is accepted in statistics, it is important to know if such error is biased or unbiased. *Biased error* (also called *systematic error* or *cumulative error*) results when the errors tend to be 'all one way', so that any resulting figures will be known to be too big or too small. Thus, if in a survey all figures are rounded down, then a biased error results since all the figures (other than any exact figures) are recorded below their true value and a total of such figures would therefore be below the true total. Another form of biased error arises when items are counted. A non-existent item is very unlikely to be counted, whereas an existing item could well be missed – so that the total count is biased towards understating the true number.

When any error is as likely to be one way as much as the other (and to much the same extent) we talk about an *unbiased error*. The error resulting from rounding to the nearest unit is, therefore, an unbiased error since any resulting figures are as likely to be overstated as they are understated.

5. Compensating errors. In many situations – and in particular when rounding to the nearest unit – an error in one figure is frequently compensated to a greater or lesser extent by an error in the opposite direction of another figure. Errors of this kind are called *compensating errors*. It is, of course, the presence of many compensating errors that causes an unbiased error to be relatively much smaller than a biased error.

Note that the term 'compensating error' is more a concept than an actual figure since it is meaningless to try and identify the original error and the subsequent compensating error. So one cannot say if a specific error is the compensating error or not. This point is, perhaps, best illustrated by the story of the lecturer who remarked that Nature always compensated – if a baby were born with its right leg longer than its left then Nature compensated by ensuring that its left leg was shorter than its right.

6. Significant figures. *Significant figures* are the digits that carry real information and are free from any inaccuracies. How many digits are in fact significant in any individual value depends on the degree of rounding that may have occurred at some earlier stage in the calculations or when the data was collected. The concept is perhaps best grasped by means of examples:

Calculated figure	Stated figure (**bold** digits) when accuracy is to:		
	Four significant figures	*Three significant figures*	*Two significant figures*
613.82	**613.8**	**614**	**610**
0.002817	0.00**2817**	0.00**282**	0.00**28**
3,572,841	**3,573**,000	**3,57**0,000	**3,6**00,000
40,000	**40,00**0	**40,0**00	**40,0**00

It should be noted that zeroes which only indicate the *place value* of the significant figures, e.g. hundreds, thousands, hundredths, thousandths, are not counted as significant digits.

7. Spurious accuracy. At this point it should be understood that it is not enough merely to be aware that complete accuracy in real-life statistics is usually impossible: it is also important that *no claim* should be made for such accuracy where it does not exist. Students may assert that they never do make such claims, but they should realise that every figure they write makes a statement regarding the accuracy of that figure.

For example, to write 4.286 means that an accuracy of up to three decimal places is being claimed. It *must* mean that, for if the accuracy is only to two decimal places then the 6 is a wild guess, in which case it is pointless to include it. But since it *is* included, a reader will assume that it does have meaning and that the writer of the figure is therefore claiming an accuracy of three decimal places. Note, incidentally, the converse – if a figure, accurate to four decimal places, comes to exactly, say, 5.3 it must be written as 5.3000 to indicate the correct level of accuracy unambiguously.

From all this it follows that a figure should include only those digits which are accurate – otherwise a greater accuracy is implied than the figure really has. When a figure implies an accuracy greater than it really has, such accuracy is termed *spurious accuracy*. Not only is spurious accuracy pointless, but it can also mislead, sometimes seriously. For instance, if a manufacturer is told that the estimated cost of an article is £266.67 and he signs a contract to sell a large quantity at £280 each, he may regret his action when he learns that the cost price was based on an estimate of £800 for making three, give or take £100.

8. Adding and subtracting with rounded numbers. When

adding or subtracting with rounded numbers it is important to remember that the answer cannot be more accurate than the least accurate figure.

Example

Add 225, 541 and 800, where the 800 has been rounded to the nearest 100.

$225 + 541 + 800 = 1,566$. But since the least accurate figure is to the nearest 100, the answer must also be given to the nearest 100, i.e. 1,600. Any attempt to be more exact can only result in spurious accuracy. Indeed, even the 6 in 1,600 is suspect, for, since 800 is subject to a possible error of 50 either way, the true answer lies between 1,516 and 1,616, i.e. 1,500 may easily be a more accurate rounded figure than 1600.

9. Multiplying and dividing with rounded numbers. When multiplying or dividing with rounded numbers one must ensure that the answer does not contain more significant figures than the number of significant figures in any of the *rounded* figures used in the calculation.

Examples

(1) *Multiply 1.62 by 3.2 (both rounded).*
$1.62 \times 3.2 = 5.184$. But since 3.2 is only two significant figures the answer can only contain two significant figures, i.e. 5.2.

(2) *Multiply 1.62 by 2, where 2 is an exact, i.e. not a rounded, figure.*
$1.62 \times 2 = 3.24$. Since 1.62 is correct to three significant figures, the answer can be left as three figures. (But note that the 4 is still suspect. The 1.62 could represent any figure between 1.615 and 1.625. These doubled are 3.23 and 3.25, either of which may be closer to the true answer than our 3.24.)

10. Maxima and minima. The above rules are what may be called the 'very least' rules, i.e. rules that at the very least must be followed. They have the advantage of being easy to remember but, while usually quite suitable in practice, they are not always free from possible slight error – as has been demonstrated. If an *error-free* figure is required then the following procedure should be followed:

(a) rewrite each rounded figure twice, first giving it the *maximum* value it could represent, and then giving it the *minimum*;

(*b*) next calculate two answers, one using the maxima and the other the minima. The true figure should be stated as lying between these two answers.

Examples

(1) *Add* 15.04, 21 *and* 10.3 (*all rounded numbers*).

Maxima	Minima
15.045	15.035
21.500	20.500
10.350	10.250
46.895	45.785

Therefore the true answer lies between 45.785 and 46.895.

This can be alternatively stated as 46.340 ± 0.555 (46.340 being the midpoint of 45.785 and 46.895).

(2) Repeat the example in **8** assuming that the first two figures have been rounded to the nearest whole unit – this time finding an error-free answer.

Maxima	Minima
225.5	224.5
541.5	540.5
850.0	750.0
1617.0	1515.0

Therefore, the error-free answer $= 1566 \pm 51$

(3) Repeat the examples in **9** to give error-free answers.

	Actual figures	Maxima	Minima	Error-free answer
1.	1.62×3.2	1.625×3.25 $= 5.28125$	1.615×3.15 $= 5.08725$	5.18425 ± 0.097
2.	1.62×2	$1.625 \times 2 = 3.25$	$1.615 \times 2 = 3.23$	3.24 ± 0.01

Note, incidentally, that the midpoint of the *maxima* and the *minima* in Example (3)1 is *not* 5.18400 (1.62×3.2) as one might have expected.

11. Absolute and relative error. The actual difference between the stated figure and the true figure is termed the *absolute error*. However, we are not often so much concerned with the absolute error as the size of that error relative to the total figure. Thus, in measuring the distance between London and Sydney an error of a kilometre is insignificant, whereas such an error in the distance between Dover and Calais could be serious. Consequently, computing the percentage that the absolute error bears to the total figure gives a measure of the *relative error*. Therefore

$$\text{Relative error} = \frac{\text{Absolute error}}{\text{Estimated figure}}$$

Example
Find the relative error in the examples in the previous paragraph:

Example	(1)	(2)	(3)1	(3)2
Absolute error	0.555	51	0.097	0.01
Estimated figure	46.34	1566	5.18425	3.24
Relative error	1.2%	3.3%	1.9%	0.3%

Note that multiplying by an exact figure (example (3)2) has a very significant effect on the relative error.

Finally, it should be appreciated that the real error in any instance is unknown – the figure actually being calculated being the *potential* error.

12. Standard error. A *standard error* is another form of statistical error that arises in the context of sampling theory (*see* Chapter 22). Its definition and significance will, then, be left until that subject is discussed.

13. Point estimate. If, when we are asked to make an estimate in circumstances when we really can only know the range within which the required value can lie, we give a single figure, then we say we're making a *point estimate* – i.e. we are answering with a single point on the scale rather than a range of values.

Progress test 2

(*Answers in Appendix 4*)

1. Explain the following forms of error:

   (*a*) statistical error (**1**);
   (*b*) biased error (**4**);
   (*c*) systematic error (**4**);
   (*d*) cumulative error (**4**);
   (*e*) unbiased error (**4**);
   (*f*) compensating error (**5**);
   (*g*) absolute error (**11**);
   (*h*) relative error (**11**).

2. Add 280 tonnes, 500 tonnes, 641 tonnes, 800 tonnes and 900 tonnes.

3. 1,200 people (to the nearest 100) each bought a sack of potatoes every quarter. Each sack weighed 65kg (to the nearest kg). Calculate the total weight of potatoes bought in the year.

4. Add the following rounded figures giving an error-free answer on the alternative assumptions that:

   (*a*) the rounding was to the nearest digit;
   (*b*) the figures were rounded up.

Figures: 2.81   4.373   9.2   5.005

Assignment

Find a table comprising about 100 figures and:

(*a*) Compute the exact total.
(*b*) Compute the total after:
   (i) rounding all figures up to the next 10;
   (ii) rounding all figures down to the previous 10;
   (iii) rounding all figures to the nearest 10.
How do the relative errors compare?
(*c*) Repeat (*b*) rounding in 100s.

Part two
The compilation and presentation of statistics

3
Collection of data

Before any statistical work can be done at all, figures must be collected. The collection of figures is a very important aspect of statistics, since any mistakes, errors or bias which arise in collection will be reflected in conclusions subsequently based on such figures.

Always remember that *a conclusion can never be better than the original figures on which it is based.* Unless the original figures are collected properly, any subsequent analysis will be, at best, a waste of time and possibly even disastrous, since it may mislead, with serious consequences.

Populations and samples

1. Populations. Before one starts collecting any data at all it is very important to know exactly what one is collecting data about! Newcomers to statistical method often fail to appreciate the importance of this; anxious to ascertain, say, local TV-viewing habits they make a rapid door-to-door survey and present their results as the TV-viewing habits of the people in their local area. However, in actual fact they did *not* collect data from the people in the area; they collected data from people in the area *who were in when they called.* If the survey had been made in the evening this factor would have obviously influenced the results. Most of the people who were usually out in the evenings would not be in and so their viewing would be unrecorded. Yet these people in the very nature of things would have different TV-viewing habits from the people whose viewing was recorded.

We call the group of people (or items) about which we want to obtain information the *population*. Clearly, the population must be defined very carefully. Thus, if we wish to investigate the colours of all the cars on English roads then our population will be all the cars on English roads. Note this is not the same as all English cars – many cars on English roads are foreign and some English cars are on roads abroad.

Defining the population may prove, in fact, very tricky. For instance, in an inquiry relating to the number of housewives who go out to work, how should 'housewife' be defined? If it is taken to mean a wife with no occupation other than housekeeping, the survey would show that *no* housewives go out to work! If it means a wife who keeps house, then the question arises whether a newlywed in a one-room flat with her husband away on business nearly all the time is really a 'housewife'. And should we include the wife who has a maid to do most of the housework? Different interpretations will lead to different results in the analysis.

Another problem sometimes associated with the population is that its full extent may not be known, e.g. the number of people in England with unsuspected diabetes. Obviously it is not easy to collect figures for such a population.

2. Samples. If our population is relatively small and easily surveyed we may well examine every item in the population. If we do this we are, incidentally, taking a *census*. However, in practice, populations are usually too big or items too inaccessible to enable the whole population to be examined. One may have to be satisfied with examining only a part, or sample, of the total population. A *sample*, then, is a *group of items taken from the population for examination.*

3. Sample frame. A list of the entire population from which items can be selected to form a sample is called a *sample frame*. Creating a sample frame often proves to be a major problem. If you wanted, for instance, to take a sample of people using a particular product it would be very difficult, if not wholly impractical, to create a list of all such people. Indeed, perhaps the best you can hope for is that the product is technical enough for it to be worth the purchaser's time to fill in a guarantee card which includes his name and address. These cards would then constitute the sample frame, although this would, of course, fail to include those purchasers who for some reason or another never filled in a card.

Sometimes the sample frame may appear obvious. Thus, the electoral roll may appear the perfect sample frame if a sample of voters in a particular town is needed. However, such rolls are often as much as a year or more out of date, and surveys have shown they are already 8–10 per cent wrong on the day they are drawn up. Not only, then, will the roll fail to be a full and accurate list of the population under investigation, but it will also contain an element of bias against the more mobile element of the population. Such voters will be slightly under-represented because the newly arrived members of this mobile class will not be on the roll while those members who have recently moved will not be available for interview. For all that, the electoral roll will almost certainly be the best sample frame obtainable.

Later in the chapter the virtue of random sampling will be explained (*see* **11**). At this point it should be appreciated that without a sample frame random sampling is impossible, and in such circumstances a non-random method of selecting a sample will need to be adopted.

4. Costs, accuracy and samples. Although to the student it may appear better to survey the whole population than rely on a small sample, this is often not so. First of all it costs considerably more to examine the whole population than just a small sample and *these higher costs could easily exceed the value of the survey results*. Thus, cost very frequently rules out examining the whole population.

Moreover, it has often been found that taking only a sample results in improved accuracy. The reason for this paradox is that a sample can be given very careful attention and measurements made with a high degree of accuracy. Examining the whole population, on the other hand, is such a major task that often unskilled investigators have to be brought in and this, coupled with the monotony of examining large numbers of items, leads to so many errors that the overall cumulative error is greater than the error inherent in using sample results to draw conclusions about the whole population.

Methods of collection

We have seen that figures relating to a chosen population can be obtained from the whole population or from a sample. Whichever approach is decided upon, one or a combination of the following methods of collection can be adopted:

(*a*) *direct observation*, e.g. counting for oneself the number of cars in a car park or examining the sales invoices of a company;

(*b*) *interviewing*, e.g. asking personally for the required information;

(*c*) *abstraction from published statistics*;

(*d*) *postal questionnaire*, e.g. sending a questionnaire by post and requesting completion.

5. Direct observation. This is the best method of collecting data as it reduces the chance of incorrect data being recorded. Unfortunately it cannot always be used, generally on account of the cost. It would be uneconomical, for instance, to follow a housewife around for a month in order to find out how many times she vacuumed the lounge, quite apart from the practical difficulties. At other times this method cannot be used because the information cannot be directly observed, e.g. where people would spend their holidays if they had unlimited money.

Probably the most fruitful context for direct observation is the collection of internal data in an organisation. Sales invoices, purchase invoices, time sheets, production records – all of these can be examined in their entirety. Of course, having access to all the data does not mean that one has completely accurate data – time sheets are notorious for recording what people want recorded rather than what actually took place.

6. Interviewing. A disadvantage of interviewing is that inaccurate or false data may be given to the interviewer. The reason may be:

(*a*) forgetfulness;

(*b*) misunderstanding the question; or

(*c*) a deliberate intent to mislead.

For example, a housewife who is asked how much milk she bought the previous week may:

(*a*) have forgotten;

(*b*) include – or fail to include – any milk bought by her husband; or

(*c*) overstate the amount because she has a number of children and feels she ought to have bought more than she did.

Another disadvantage of interviewing is that if a number of interviewers are employed they may not record the answers in the same way as the investigator himself would have done. For example, to the question 'Did you watch the XYZ TV programme last night – yes or no?' the interviewer may get the answer, 'Well, part of it, then someone called and we switched off.' The interviewer may well record this as a 'no' answer, while the investigator may be working on the assumption that such answers are being recorded as 'yes'.

Different standards like this can easily result in the wrong conclusions being drawn from the survey. One way of overcoming this disadvantage is to train the interviewers very carefully so that they record data in *exactly* the same form as the investigator himself would record it.

7. Abstraction from published statistics. Any data that an investigator collects himself is termed *primary data* and, because he knows the conditions under which it was collected, he is aware of any limitations it may contain.

Data taken from other people's figures, on the other hand, is termed *secondary data*. Users of secondary data cannot have as thorough an understanding of the background as the original investigator, and so may be unaware of such limitations.

Statistics compiled from secondary data are termed *secondary statistics*. Obviously the compilation of such statistics needs care, in view of the possibility of there being special features concerning the earlier statistics, or the population concerned, which are not known to the compiler. For example, a table relating to remuneration may cover only basic pay; if this fact were not indicated, a subsequent investigation using the figures in conjunction with a survey on total earnings (e.g. including overtime) could result in quite false conclusions.

For this reason anyone wishing to use published statistics should consider the purpose for which they were originally compiled. In many government publications the statistics are compiled with the expectation that they will be used in the production of secondary statistics. Such statistics are carefully annotated and explained so that users will not be misled, and secondary statistics may be prepared from them with reasonable confidence. However, many others are published in connection with a specific inquiry and it may be very dangerous to use them as a base for the compilation of secondary statistics.

8. Postal questionnaire. This is the least satisfactory method, for the simple reason that only relatively few such questionnaires are ever returned. A return of 15 per cent is often considered a good response for certain types of survey, although reminder notices can usually improve the percentage. Moreover, the questionnaires which are returned are often of little value as a sample, since they are frequently biased in one direction or another. For example, a questionnaire relating to washing machine performance will be returned mainly by people with complaints who are only too pleased at the chance to air them. Satisfied users will probably not bother to reply. A conclusion, based on the returned questionnaires, that washing machine performance was on the whole bad would therefore be a false conclusion.

Bias of this sort is rarely as obvious as in the above example and so cannot be allowed for in any analysis of returned questionnaires. Postal questionnaires therefore are not recommended unless one of the following conditions applies:

(*a*) completion is a legal obligation, e.g. government surveys;

(*b*) all the non-responders are subsequently interviewed in order to obtain the required information;

(*c*) an appropriate sample of the non-responders is interviewed and it is established that the failure to reply *is in no way connected with any bias.*

9. Design of a questionnaire. If a questionnaire is to be used, either as a postal questionnaire or as a basis for interviewing, the following points should be observed in its design.

(*a*) Questions should be *short* and *simple*. It is far better that there are many short questions than that there are a few long questions.

(*b*) Questions should be *unambiguous*.

(*c*) The best kinds of question are those which allow a *preprinted answer* to be ticked.

(*d*) The questionnaire should be as *short* as possible.

(*e*) Questions should be *neither irrelevant nor too personal.*

(*f*) *Leading questions should not be asked*. A 'leading question' is one that suggests the answer, e.g. the question 'Don't you agree that all sensible people use XYZ soap?' suggests the answer 'yes'.

(*g*) The questionnaire should be designed so that the *questions fall into a logical sequence*. This will enable the respondent to understand its purpose, and as a result the quality of his answers may be improved.

10. Problem of extreme values. When reviewing collected data it may well happen that a rather surprising extreme value is noticed. Thus, if in data about ages of infants in a school we find a figure of 54 years, it suggests that one of the teacher's ages was collected by mistake. Similarly, a height of 62 feet in data relating to men's heights suggests a misreading of the scale or a clerical error. In these cases the figures are so obviously erroneous that they will be removed from the data. However, it is often not clear whether a figure is an error or is genuine, and so there is some doubt as to what should be done. To include an erroneous figure in the data would distort the results, but to ignore a genuine figure would also cause distortion.

The final decision must, of course, be left to the investigator's judgement and will depend on the individual circumstances. However, wrongly including an extreme value is usually regarded as risking a more serious error than wrongly ignoring one. After all, statistics is about realistic patterns and an extreme value, even if correct, will probably be so untypical as to tend to mislead any user of the information to a greater or lesser extent. To say, for instance, that the average capital owned by the occupants of a shanty town is £10,000 when, but for an eccentric millionaire in their midst, nobody owns more than £500 is hardly likely to be a useful piece of information. So, while ignoring an extreme value may seem a little dishonest, it may nevertheless result in a more accurate *impression*.

Random samples

Taking a sample is not simply a matter of taking the nearest item. If worthwhile conclusions relating to the whole population are to be made from the sample, it is essential to ensure as far as possible that the sample is free from *bias*, which can be defined as the influence in the selection of a sample of a particular feature in excess of its true importance.

Assume, for example, that we wish to know what proportions of Europeans are fair-haired. If we took a sample wholly from Stockholm our conclusion based on it would be wrong, because Swedes are a fair-haired nation. Our sample would thus allow 'fair-hairedness' to have an importance greater than is warranted, i.e. it would be biased towards 'fair-hairedness'.

Unfortunately it is not sufficient merely to ensure there is no *known* bias in our sample. *Unsuspected* bias can equally invalidate our conclusions. So the question arises how one can select a sample that is free even from unknown bias.

11. What is a random sample? The possibility of taking a sample having unsuspected bias can be reduced by taking a random sample. A *random sample* is a sample *selected in such a way that every item in the population has an equal chance of being included*. This is the only method of sampling in which we can be confident that the selection method is free from bias.

There are various methods of obtaining a random sample but they all depend on the selection being wholly determined by chance.

One may imagine such a selection being made by writing the name (or number) of each item in the population on a slip of paper and then drawing from a hat, as in a lottery, the required number of slips to make up the sample. Although this gives the idea behind 'random selection', in practice great care has to be taken that no bias can possibly arise, e.g. in this method adjacent slips could conceivably stick together. Probably the best method of selection is to number all the items in the sample frame and then allow a computer to generate a series of random numbers which will identify the items to be selected for the sample.

While the use of a computer may be the easiest way of selecting a genuine random sample, in practice a table of random numbers may be used instead. Such a table is simply published as a long sequence of random digits set out in blocks for ease of reading. The application of these tables to sample selection can be illustrated as follows:

Example

A sequence of published random numbers runs as follows:

54261 90067 02374 82816 39210 73829

The sample frame for the survey involved shows a total of 642 items and a random sample of 6 items is required.

This sample can be selected by dividing the random digits into sets of three and selecting the first six items thus indicated, ignoring any sets with values above 642, i.e.

542/61 9/006/7 02/374/ 828/16 3/921/0 73/829

Thus numbers 702, 828 and 921 would be ignored, leaving items 542, 619, 6, 374, 163 and 73 comprising the random sample.

12. Random samples are not perfect samples. Finally, it must be clearly appreciated that a random sample is not necessarily a good cross-section of the population.

Drawing the names of Europeans out of a hat *could* result in a sample containing all Swedes – though it is highly unlikely. Thus a random sample, too, can be one-sided and does not guarantee a *sample* free from bias. It simply guarantees that the *method of selection* is free from bias. This is a rather subtle difference, but an important one.

Other sampling methods

There are often occasions when the selection of a pure random sample is not feasible. These occasions arise:

(*a*) when such a sample would entail much expensive travelling for the interviewers;

(*b*) when 'hunting out' the people selected would be a long and uneconomic task;

(*c*) when there is no sample frame – one cannot select a random sample of fair-haired mothers, for example, since there is no sample frame for these mothers.

These obstacles are overcome by using *multi-stage, quota* and *cluster* sampling respectively. Additionally, there is a 'short-cut' selection method, *systematic sampling*, and a method, *stratified sampling*, which, exceptionally, is an improvement on simple random sampling.

13. Multi-stage sampling. The objective of *multi-stage sampling* is purely and simply to save time and money. If a nationwide survey is needed and a sample frame exists (e.g. the electoral roll), there is no theoretical objection to taking a pure random sample. However, almost certainly such a sample will be scattered the length and breadth of the country and the time and expense of sending interviewers to each selected person would be prohibitive.

To bring the survey within practical and economic bounds multi-stage sampling is adopted. In this method the country is first divided into a number of large areas (e.g. counties) a few of which are then selected at random. These areas are in turn divided into a number of smaller areas (e.g. parishes) and a few of these also selected at random. This procedure is continued until the areas selected are small enough for individuals to be selected at random (e.g. streets or 1 cm. ordnance survey squares), and these individuals collectively form the sample. Note that at each stage either the size of the sample taken within each area, or each area's probability of selection, must be in proportion to the population of that area, otherwise there will be a bias towards areas of low population. (If this point is not clear imagine you are to take a sample of one individual from two areas, the first containing 19 men and the second one woman. If one of the areas is selected at random, and then an individual within that area is selected at random, there is a 50/50 chance the area with the one woman will be selected and a certain chance she will be the individual selected. So the initial probability of selecting a woman is 1 in 2 whereas it should be 1 in 20. This form of bias can be redressed by arranging that the first area has a 19 in 20 chance of being selected, and the second 1 in 20.)

It is a feature of this method that every person initially has an equal chance of being selected – as in the case of a pure random sample. However, the method ensures that at the end the actual people selected will be concentrated in specific areas. Since neighbours tend to be similar to each other – either ethnically, economically, or in respect of occupation – some bias in these respects is virtually inevitable in the final result.

14. Cluster sampling. *Cluster sampling* is akin to multi-stage sampling in so far as the country is again repeatedly subdivided into areas which are selected at random – and individuals selected at random from the final areas. However, cluster sampling is adopted when there is no sample frame from which the final sample can be selected. In this type of sampling the interviewer has to comb his area meticulously to find the items needed to form that particular area's sample.

Assume, for instance, we needed to form a sample of fair-haired mothers or adult oak trees. First, a number of small areas are selected in the same way as in multi-stage sampling. The size of these areas will to a large extent depend upon the believed density of the items to be sampled, e.g. areas will be small in the cities for the mothers, but large for the oak trees. An interviewer then visits each area with the objective of finding every fair-haired mother or adult oak tree that he can.

Cluster sampling has the same disadvantage as multi-stage sampling – that since the items in the area will tend to resemble each other, some bias will result. Additionally, in cluster sampling there will be a tendency for the less obvious items to be missed. Since it may be that reclusive fair-haired mothers, or oaks hidden by other trees, have features a little different from their more obtrusive counterparts, another avenue for bias arises. However, in the absence of a sample frame, cluster sampling is often the only viable method of selecting a sample at all.

15. Quota sampling. Another method of selecting a sample in the absence of a sample frame or where the cost of a random sample is prohibitive is by *quota sampling*. This method is similar to cluster sampling except that the interviewer does not have to find all the items with a given characteristic but only a predetermined number which is termed his *quota*. Once he has filled his quota his work is finished.

In practice an interviewer's quota is very often subdivided into classes. Thus, 20 per cent of his quota may be people 20–30 years of age, 50 per cent 30–40, and 30 per cent over 40. Also, his quota can be subdivided into economic classes from the upper class down to the unskilled worker. All these subdivisions are designed so that the final sample comprises as accurate a cross-section of the population being sampled as possible.

Although quota sampling is relatively very cheap, it does have the disadvantage of tending to be rather imprecise. Very much depends on the interviewer's skill in weighing up into just which classification a potential interviewee falls. As this is often very subjective there is plenty of room for divergent judgement. Moreover, much will also depend on where the interviewer looks to find his quota. A sample of shoppers selected solely from among the shoppers on the concourse of a railway terminal will obviously be biased – local shoppers will hardly be represented at all. This is in addition to the bias which inevitably results from the fact that people who refuse to be interviewed may well have some special characteristic which is relevant to the survey. As can be imagined from all this, where quota sampling is to be used it is vital that the interviewers are well trained so that all the potential forms of bias are minimised as far as is possible.

16. Systematic sampling. *Systematic sampling* is simply a short-cut method for obtaining a virtually random sample. If a 10 per cent sample (say) is required, then the sample can be selected by taking every tenth item in the sample frame, provided there is no regularity within the frame such that items ten spaces apart have some special quality. If by mischance there happened to be some pattern in the frame that coincided with the sampling interval, the sample would be extremely biased, e.g. every tenth house in a street might have a bay window and therefore be slightly more expensive. Such bias is not common, however, and systematic sampling can often be safely used.

17. Stratified sampling. It is important to note that none of the techniques so far mentioned is better than a pure random sample. *Stratified sampling*, however, is *better* than purely random methods and must therefore be distinguished from the others. In order to use it, however, you have to know what groups comprise the total population, and in what proportions (it is, in fact, because you are able to inject such supplementary information into your normal random sampling technique that it is possible to obtain the improved results that emerge).

For instance, assume that in a survey relating to wages in a particular industry there are male and female workers. As the proportion of each can easily be found, stratified sampling can be employed. The technique involves the following steps (assuming in this case that the relevant proportions are 3:7):

(*a*) decide on the total sample size (say 1,000);

(*b*) divide this into subsamples with the same proportions as the groups in the population (300 and 700):

(*c*) select at random from within each group (stratum) the appropriate subsample (300 male and 700 female workers);

(*d*) add the subsample results together to obtain the figures for the overall sample.

One reason why stratified sampling is an improvement over a pure random sample is that it lessens the possibility of one-sidedness. As we have seen, a random sample of Europeans *could* be composed wholly of Swedes. But if it were arranged that the sample should contain different proportions of each nationality according to the size of the country's population, such a one-sided sample would be impossible.

Although the theory is outside the range of this book it may be noted that step (*b*) can be modified so that the subsamples are proportional to the variability within the strata (as measured by their standard deviations – *see* 11:**5**) though an adjustment for this modification needs to be made at step (*d*). To see broadly why this modification is useful imagine that the 300 male employees all receive exactly the same wage (i.e. the variability is nil). Clearly, in such a case we need only take a sample of one from this stratum. This means we could then have a subsample of 999 females from the second stratum which, of course, will mean that our final answers will be just that much more precise.

Progress test 3

(*Answers in Appendix 4*)

1. Distinguish between:

(*a*) (*i*) population (**1**);
  (*ii*) sample (**2**);

(*b*) (*i*) primary data (**7**);
  (*ii*) secondary data (**7**);

(*c*) (*i*) multi-stage sampling (**13**);
  (*ii*) cluster sampling (**14**);
  (*iii*) quota sampling (**15**);
  (*iv*) systematic sampling (**16**);
  (*v*) stratified sampling (**17**).

2. Define carefully:

(a) sample frame (**3**);
(b) random sample (**11**).

3. What comments have you to make on the following statement made in *Mack's Mag?*

'We thought we would like to learn something of the physical characteristics of our readers so ten newsagents were selected and an observer stationed at each. This observer noted down physical facts about every person who bought a copy of *Mack's Mag* and the results of this random sample will be published in next week's issue.'

4. What sampling methods would you use to obtain the following information?

(a) Ages of Australian-born persons resident in the UK.
(b) Health details relating to UK borough councillors.
(c) Cinema takings.
(d) The views of the public on Sunday trading.

5. A social research group plans to survey the telephone subscribers in a particular telephone area. They select a sample of these subscribers by taking the first 200 names in the area telephone directory. What bias is such a selection method likely to introduce?

Assignments

1. Take a random sample of 500 letters from words in this book. Then take 100 pages at random and record the first letter on the page. Can you detect any sign of bias in treating these first letters as a sample of all the letters in the book? Repeat this taking the last letter of 100 pages.

2. Can you devise a non-random method of taking a sample of 100 letters which is, as far as you can ascertain, free from bias?

4
Tables

It is a psychological fact that data presented higgledy-piggledy is far harder to understand than data presented in a clear and orderly manner. Consequently, the next step after the figures have been collected is to lay them out in an orderly way so that they are more readily comprehended.

1. Tables. Since a piece of paper is two-dimensional, the most effective layout is almost always one of columns and rows. Such a layout is termed a *table*. To illustrate the superiority of a table over, say, a simple narrative statement of figures look at the following two methods of presenting the same data:

(*a*) *Narrative form.* Sales of tables and chairs at Mr Fred's factory were analysed from his sales invoices for the years 19−1 to 19−5. Unfortunately, owing to the somewhat idiosyncratic style of Mr Fred's bookkeeping there arose some ambiguity as to the exact number of sales. In the circumstances it was felt that accuracy could only be guaranteed to the nearest thousand, and here the figures have all been rounded to this degree of accuracy.

In 19−1 sales to the Western region amounted to 4,000 tables and 8,000 chairs and to the Eastern region 8,000 tables and 16,000 chairs. Of the tables 4,000 red, white and blue were each sold while the chair colours were 6,000 red, 12,000 white and 6,000 blue. The following year, 19−2, Western region sales were 6,000 tables and 10,000 chairs and Eastern region 8,000 tables and 14,000 chairs. Broken down by colour the sales were 2,000 red, 6,000 white and 6,000 blue tables and 8,000 red, 10,000 white and 6,000 blue chairs. Sales in 19−3 were 8,000 tables and 16,000 chairs for the Western region and 6,000 tables and 16,000 chairs for the Eastern region. Colours for this year were 6,000 white and 8,000 blue tables – with no red tables sold – and 4,000 red, 18,000 white and 10,000 blue chairs. Parallel figures for 19–4 and 19–5 were: Western region, 12,000 and 22,000 tables, and 20,000 chairs in each year, and the Eastern region 12,000 tables each year and 24,000 and 28,000 chairs. As to colours, these were 6,000 red, 6,000 white, 12,000 blue, and 8,000 red, 10,000 white, 16,000 blue tables in each year respectively, while the chair colours were 10,000 red, 20,000 white, 14,000 blue, and 8,000 red, 26,000 white and 14,000 blue respectively. The total sales of tables for the years 19–1 to 19–5 were 12,000, 14,000, 14,000, 24,000 and 34,000, and the total sales of chairs were 24,000, 24,000, 32,000, 44,000 and 48,000.

(b) *Tabular form.*

Table 4A   *Sales of tables and chairs – Fred's factory 19–1 to 19–5 (000s)*
Sales analysed by region and colour   T = Tables; C = Chairs

	19–1		19–2		19–3		19–4		19–5	
Region	T	C	T	C	T	C	T	C	T	C
Western	4	8	6	10	8	16	12	20	22	20
Eastern	8	16	8	14	6	16	12	24	12	28
Total	12	24	14	24	14	32	24	44	34	48
Colour										
Red	4	6	2	8	0	4	6	10	8	8
White	4	12	6	10	6	18	6	20	10	26
Blue	4	6	6	6	8	10	12	14	16	14
Total	12	24	14	24	14	32	24	44	34	48

Source: Fred's sales invoices, 19–1 to 19–5.
Footnote: Owing to the somewhat idiosyncratic style of Mr Fred's bookkeeping there arose some ambiguity as to the exact numbers of sales. In the circumstances it was felt that accuracy could only be guaranteed to the nearest thousand and here the figures have all been rounded to this degree of accuracy.

While a table is obviously far clearer than a narrative form of presentation its construction does call for considerable care.

2. The prime requirement of table construction. The construction of a table is in many ways a work of art. It is not enough just to have columns and rows: a badly constructed table can be as confusing as a mass of data presented in narrative form. Yet, as with a work of art, it is difficult to lay down precise rules that will apply to all cases. For this reason the reader should construct his tables as common sense guides him, and the sounder his common sense, the better his tables will be.

On the other hand there are certain principles which must be observed in the construction of *all* tables. Though imagination can often improve a table, to ignore any of these principles will only result in a loss of clarity and impact which might be compared with mumbling instead of speaking clearly.

3. Basic principle of table construction. Of all the principles of table construction, there is one which is basic: *construct the table so that it achieves its object in the best manner possible.* This means the person constructing the table must ask himself at the very beginning, 'What is the purpose of this table?' Some of the possible reasons for which a table may be constructed are:

  (*a*) to present the original figures in an orderly manner;
  (*b*) to show a distinct pattern in the figures;
  (*c*) to summarise the figures;
  (*d*) to publish salient figures which other people may use in future statistical studies. (Many government statistics are produced for this purpose, and the reader is warned against using such tables as models if his own table is produced for a different purpose.)

A major issue in table construction is deciding which columns of figures should be adjacent to each other. For example, would Table 4A be improved if there were two main columns, 'Tables' and 'Chairs', each subdivided into the five years? The answer depends, as always, on the purpose of the table. If sales of tables and chairs are to be compared, it is best as it stands. But if a comparison between sales year by year is required then the change would be an improvement.

Other kinds of decisions must also be made. For instance, what totals should be shown? Should percentages be included? Should some figures be combined, or even eliminated? Invariably, the basis of all these decisions is *the purpose to which the table will be put.*

4. Other principles of table construction. While the basic principle of table construction should always be paramount, the following additional principles should also be observed.

(*a*) *The table should be simple.* This is vitally important. A table with too much detail or which is too complex is much harder to understand, and so defeats its own object. *Remember*, it is better to show only a little and have that understood than to show all and have nothing understood.

(*b*) *The table must have a comprehensive, explanatory title.* If such a title would be too long, then a shorter one may be used together with a subtitle.

(*c*) *The source must be stated.* All figures come from somewhere and a statement of the source must be given, usually as a footnote.

(*d*) *Units must be clearly stated.* It is possible to reduce the number of figures by indicating in the title, on the top right of the table, or in the column headings the number of thousands, or multiples of ten, each figure represents (e.g. writing £000 in the title indicates that all figures in the table are in thousands of pounds).

(*e*) *The headings to columns and rows should be unambiguous.* It is very important there should be no doubt about the meaning of a heading. If a lengthy heading would be necessary to remove ambiguity, then a short heading may be used with a symbol referring the reader to a footnote containing a more detailed explanation.

(*f*) *Double-counting should be avoided.* If a table shows *People in X: 100*, and also *People in X and Y: 500*, then the 100 people in X appear twice in the table. This is 'double-counting' and, as it is apt to mislead, should normally be avoided. Cumulative figures may, of course, be shown but should be identified as such.

(*g*) *Totals should be shown where appropriate.* Totals are used in a table for one of the following purposes:

   (*i*) to give the overall total of a main class;

   (*ii*) to indicate that preceding figures are subdivisions of the total;

   (*iii*) to indicate that all items have been accounted for, e.g. if a survey of 3,215 people is presented in a table, 'Total 3,215' indicates that all the people surveyed appear in the table, i.e. there are no gaps.

   This is particularly useful if the same data has been analysed in two different ways since the two identical analysis totals help confirm that the data has, in fact, been subject to such dual analysis (*see* Table 4A).

(*h*) *Distinctive rulings should be used as appropriate.* Distinctive rulings (double lines, heavy single lines) enable different areas of the table to be relatively isolated from each other so that the user's eye is drawn to the figures that the person constructing the table wants the user to compare.

(*i*) *Footnotes should be used to qualify or clarify the table.* More often than not the figures in a table are influenced by some factor which is not discernible from the table itself (in the case of Table 4A, that the original figures were only accurate to the nearest thousand). In such a case it is essential to append a footnote to the table drawing the user's attention to this factor.

5. Derived statistics. Frequently figures in tables become more meaningful if they are expressed as percentages or, less often, as ratios. Such figures are called *derived statistics* and in constructing a table it is important to decide whether or not it can be improved by including such figures. If it can, then additional columns should be inserted in the table and the derived statistics computed and entered.

6. Advantages of tabular layout. Comparison of the narrative and tabular forms as presented in **1** shows that the tabular form of presentation has several distinct advantages, quite apart from being more readily intelligible. Primarily, these are:

(*a*) it enables any desired figures to be located more quickly;

(*b*) it enables comparisons between different categories to be made more easily;

(*c*) it reveals patterns within the figures which cannot be seen in the narrative form, e.g. Table 4A reveals that the most popular chair colour is white;

(*d*) it takes up less space, or it is far less dense.

7. Table overview. In constructing a table one can often become so immersed in the detail that errors of the wider kind are unnoticed. It is essential, therefore, that after the table is completed it is looked at as far as possible with fresh eyes. For instance, in one part of the table figures may be given to one decimal place. This implies that *all* the figures in the table are accurate to one decimal place. Consequently, a figure of, say, 34 which has been rounded will be read as 34.0 and so result in spurious accuracy. In such a case the decimal figures must be rounded to the nearest unit (or the two levels of accuracy in the table made very clear). Overviews also enable one to think carefully about all statements made in the table and avoid, for instance, titles such as 'County court judges broken down by age and sex'.

Progress test 4

(*Answers in Appendix 4*)

1. A section of roadworks where traffic is reduced to a single lane, with directions alternating, is controlled by traffic lights with two cycle time settings – 100 seconds and 160 seconds. Whichever setting is used both ends together must show STOP for 20 seconds to allow the traffic in the lane to clear before the traffic coming in the opposite direction is allowed to move. This means that during the 100-second cycle the lights in each direction are at GO for 30 seconds and at STOP for 70 seconds, and during the 160-second cycle at GO for 60 seconds and at STOP for 100 seconds. When the lights are at GO vehicles in the queue enter the lane at the rate of one every one and a half seconds.

The traffic engineer is not sure which is the best setting so he arranges an experiment in which the queue length is recorded both:

(*a*) just as the lights change from STOP to GO;

(*b*) just as the lights change from GO to STOP (i.e. the queue of vehicles which fail to enter the lane during the GO phase).

He runs this experiment for 100 cycles at each of the two settings when the average traffic arrival rate is 9 vehicles per minute, 12 vehicles per minute and 15 vehicles per minute (i.e. six 100-cycles). Since the traffic arrives purely at random (e.g. since 12 vehicles a minute means 0.2 vehicles per second, at each second there is a 0.2 probability – or 1 in 5 – of a vehicle arriving) the queue lengths vary from cycle to cycle.

The engineer wishes to have a table which shows for each part of the experiment the average and maximum queue lengths at each of the light changes for each of the six 100-cycle experiments and also, as far as possible, the average waiting time per vehicle in each experiment.

Design the table and, as well as you are able, insert illustrative figures (though this latter requirement is incidental to the table design), and comment on this data.

2. Criticise the following table:

Castings	Weight of metal	Foundry hours
Up to 4 kg	60	210
Up to 10 kg	100	640
All greater weights	110	800
Others	20	65
	290	2000

Assignment

Obtain as much data as is available about the students in your college (age, sex, subjects taken, type of residence, etc.) and present this data in informative tables.

5
Graphs

Tables, as we have seen, make data easier to understand. A further gain in this respect can often be obtained by representing the data *visually*. Such a gain stems from a further psychological fact – that people, not being computers, are able to see spatial relationships much better than numerical relationships.

For example, in comparing sales of A with sales of B in the following table, what conclusions can you draw?

	19–1	19–2	19–3	19–4	19–5	19–6
Sales of A (units)	1121	1233	1356	1492	1641	1805
Sales of B (units)	292	321	353	387	428	470

It takes a certain amount of study to see that sales of A increase each year by more units than sales of B. But this is obvious at once when the same data is shown visually (*see* Fig. 5.1).

Such visual presentation can take many forms. In this chapter we will look at graphical presentation.

Introduction

Before starting to look at graphical presentation, four definitions should be noted.

1. Graph. A *graph* is the *representation of data by a continuous curve on ruled paper*. The essence of a graph is that measurements are significant in both dimensions – i.e. the vertical height of the curve measures one variable while the horizontal distance measures another.

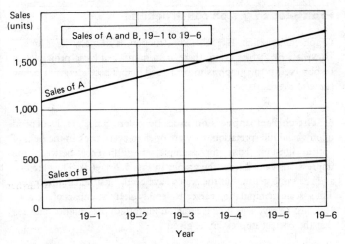

Figure 5.1 *Graph as a means of comparing sales*

2. Diagram. In the next chapter we will look at diagrams. At this point, however, it is worthwhile contrasting a diagram with a graph. A *diagram* can be defined as *any two-dimensional form of representation in which only one variable is depicted.* Usually the variable is measured by height – i.e. the height of the diagram measures whatever figure is to be presented – although sometimes angle size is used (as in pie charts, *see* 6:**9** *et seq.*). Diagrams, therefore, can be employed to represent one-dimensional data which, by definition, cannot be graphed (e.g. you cannot draw a graph showing the number of unemployed in each of the EEC countries).

3. Curves. When discussing graphs it should be understood that *any* line on a graph that represents the data to be presented is called a *curve*, even if it is a straight line.

4. Linear scale graph. In by far the majority of cases the horizontal and vertical scales are like the scales on a ruler, i.e. where, say, six units is represented by the same span on all parts of the scale. Such a scale is called a *linear* (or *normal*) *scale*, and in all but the last section of this chapter we will be looking at linear scale graphs. In the last section, however, we will look at other kinds of scales.

Principles of graph construction

Graph construction, like table construction, is in many ways an art. However, like tables again, there are a number of basic principles to be observed if the graph is to be a good one. These are given below (*see also* Fig. 5.3).

5. The correct impression must be given. Since graphs depend upon visual interpretation, they are open to every trick in the field of optical illusion. Note for example, the difference between the impressions gained from the graphs in Fig. 5.2(*a*)–(*d*). They are one and the same graph, but the scales have been constructed differently. Thus, scale manipulation can considerably alter the impact of a graph. Needless to say, *good* (i.e. accurate, undistorted) presentation ensures that the correct impression is given.

6. The graph must have a clear and comprehensive title.

7. The independent variable should always be placed on the horizontal axis. When starting a graph the question always arises which variable should be placed on the horizontal axis and which on the vertical. Careful examination will generally show that the figures relating to one variable would be quite unaffected by changes in the other variable. The variable that will not be affected is called the *independent* variable and should be placed on the horizontal axis. Note that chronological time is *always* the independent variable and so is always on the horizontal axis.

8. The vertical scale should always start at zero. Again, this is done to avoid giving wrong impressions (*see* Fig. 5.2(*a*) and (*c*)). If it is not practical to have the whole of the vertical scale running from zero to the highest required figure, then the scale may be such that it covers only the relevant figures *provided* that zero is shown at the bottom of the scale and a *definite break in the scale is shown* (*see* Fig. 5.3).

9. A double vertical scale should be used where appropriate. If it is desired to show two curves which normally would lie very far

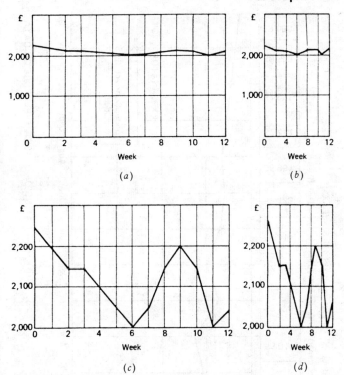

Figure 5.2 *Effect of using different scales for the same graph*

apart, two scales may be put on the vertical axis, one curve being plotted against one scale and the other curve against the second scale. Figure 5.4 is an example of a graph with a double vertical scale.

10. Axes should be clearly labelled. Labels should clearly state both (*a*) the *variable* and (*b*) the *units*, e.g. 'Distance' and 'Kilometres', 'Sales' and '£s', 'People viewing TV' and 'Thousands'.

11. Curves must be distinct. The purpose of a graph is to emphasise pattern or trend. This means that curves must be distinct.

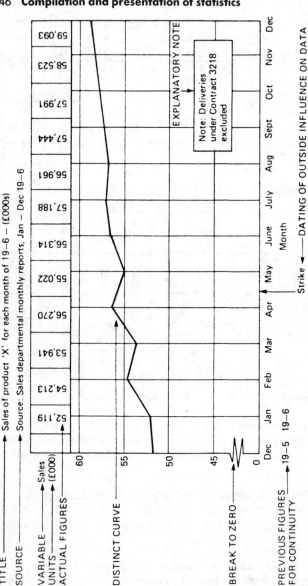

Figure 5.3 Model graph

If two or more curves are graphed, there must be no possibility of the curves being confused. To avoid such a possibility colour can be used to distinguish the curves. Alternatively, curves may be depicted as different kinds of dotted lines. Where there are two or more curves it is important, of course, that it must be very clear which data each curve represents.

12. The graph must not be overcrowded with curves. Too many curves on a graph make it difficult to see the pattern formed by any one curve and the whole point of graphical presentation is lost. How many is 'too many' depends on circumstances. If the curves are close together and intersect, the limit may easily be three or even two. Where they are well spaced and do not intersect, many more may be put on the same graph.

13. The source of the data must always be given. The source of the graphed data must always be given so that the user of the graph can, if he wishes, refer to the actual figures on which the graph is based. Sometimes the actual figures are inserted on the graph alongside the plotted points. Alternatively, they can be written at the top of the graph immediately over the point to which they relate (*see* Fig. 5.3).

14. Presentational and mathematical graphs. It should be appreciated that there are two different reasons for preparing graphs and this leads to two different kinds of graphs. These are:

(*a*) *Presentational graphs.* Presentational graphs are graphs used to *present information,* i.e. the type of graph we are concerned with in this chapter. Two points arise with this form of graph:

(*i*) *Curve thickness.* Since the object of a presentational graph is to depict information in a clear visual way, all curves should be very distinct. This means they will almost certainly be too thick for values to be read from them. But this is of no account since the purpose is to give an impression, not detailed figures. If the user wants the detailed figures he can always refer to the source. Alternatively, the actual figures can be inserted alongside the plotted points or even at the top of the graph over the individual points to which they relate (*see* Fig. 5.3).

(*ii*) *Time series plots.* When plotting points on a time series graph (a graph with chronological time along the horizontal axis), then:

(1) plot *totals* at the *end* of the periods to which they relate;

(2) plot *averages* at the *midpoint* of the periods to which they relate.

Note, incidentally, that on a time series chart a time period is the span between two vertical ordinates. On presentational graphs, however, the time period is often marked against the vertical ordinate (*see* Fig. 5.3). This really indicates the time period end, so the vertical ordinate above 'Jan' in Fig. 5.3 is the *end* of January.

(*b*) *Mathematical graphs.* Mathematical graphs are graphs drawn so that previously unknown figures can be read from them (e.g. the regression lines and the time series graphs looked at in Chapters 13 and 17 respectively). Note how these graphs differ from presentational graphs in the following respects:

(*i*) *Curve thickness.* Since values are to be read off a mathematical graph it is important that the curve be as thin as possible so that the values can be read with accuracy.

(*ii*) *Time series plots.* When plotting points on a time series graph it is vitally important that *all figures are plotted at the midpoint of the periods to which they relate*. It follows, too, that the periods should be labelled on the horizontal axis *between* the vertical ordinates, not below the ordinates as in Figure 5.3.

If you are uncertain as to which kind of graph you are dealing with, ask yourself if the graph will be used to find some *otherwise unknown value* or not. If it will be, it is a mathematical graph.

Moving totals and moving averages

When we graph a time series the actual figures for the individual periods often do not enable us to appreciate just how the series as a whole is developing. In this section we look at a simple technique that helps us to overcome this difficulty.

15. The inadequacy of a 12-month series. Look at the following table. Do you think business is improving?

Table 5A   *Monthly sales: Fred's retail shop 19–6*

Month	Jan	Feb	Mar	Apr	May	June
Sales (£)	4,000	4,100	4,200	4,300	4,400	4,500
Month	July	Aug	Sept	Oct	Nov	Dec
Sales (£)	4,600	4,700	4,800	4,900	5,000	5,100

On the face of it, business seems to be improving steadily. But, what if the sales for the previous year were as follows?

Month	Jan	Feb	Mar	Apr	May	June
Sales (£)	5,000	5,200	5,400	5,600	5,800	6,000
Month	July	Aug	Sept	Oct	Nov	Dec
Sales (£)	6,200	6,400	6,600	6,800	7,000	7,200

Clearly, business is *not* improving. Sales in January 19–6 were £1,000 below those of the previous January, and each month the gap between the sales for 19–6 and the same month the year before increases, until by December the difference is £2,100. So it can be seen that, although each month in 19–6 is better than the months before, it is worse – and progressively worse – than the same month in the previous year.

16. Moving total. Obviously the figures given at first were misleading, and a method which avoids this sort of wrong impression would clearly be useful. Direct comparison of the figures for one month with the same month the previous year is a possible solution, but this sort of comparison does not allow an overall trend to be easily observed. A better solution is the use of a *moving value*.

Examination of Fred's retail shop sales figures shows that the business is seasonal – indeed, it was for this reason that the 19–6 figures on their own gave a false impression. This problem of seasonal influence frequently arises in a time series and an excellent method of eliminating such influences is to *add together twelve consecutive months*.

Such a total is inevitably free of any seasonal influence since all the seasons, busy and slack, are included in the total. So, if we add the twelve months immediately preceding the end of *each and every* month in the table we shall obtain a series of totals, one for each month, and each total will be the total for the year immediately preceding the end of that month. Such a series is called a *moving total* or – more specifically in this case, since the totals are yearly totals – a *moving annual total* (MAT for short).

Note: Totals need not be yearly totals: they can relate to any period of time. There are, for instance, 5-year and 10-year moving totals.

17. Calculation of a moving total. Calculating a moving total is simply a matter of adding the figures for the appropriate group of periods immediately preceding the end of each individual period. However, the actual computing work can be reduced if it is appreciated that, once the first total has been found, the next total will be the same except for the difference between the new period which has been added and the old period which has been dropped, i.e. previous month's MAT minus this month's figure last year plus this month's figure this year. Table 5B demonstrates this method of calculating the moving annual total for sales at Fred's retail shop.

Table 5B   *MAT of Fred's retail shop sales*

Year	Month	Sales (£)	MAT (£)	Notes on calculation of MAT
19–5	January	5,000		
	February	5,200		
	March	5,400		
	April	5,600		
	May	5,800		There can be no MAT
	June	6,000		until 12 months'
	July	6,200		figures are available
	August	6,400		
	September	6,600		
	October	6,800		
	November	7,000		
	December	7,200	73,200	Total of sales Jan–Dec 19–5
19–6	January	4,000	72,200	73,200 minus Jan 19–5 plus Jan 19–6
	February	4,100	71,100	72,200 minus Feb 19–5 plus Feb 19–6
	March	4,200	69,900	71,100 minus 5,400 plus 4,200 etc.
	April	4,300	68,600	
	May	4,400	67,200	
	June	4,500	65,700	

Year	Month	Sales (£)	MAT (£)	Notes on calculation of Mat
	July	4,600	64,100	
	August	4,700	62,400	
	September	4,800	60,600	
	October	4,900	58,700	Add Jan–Dec 19–6 to
	November	5,000	56,700	cross-check the accuracy
	December	5,100	54,600	of this final figure

18. Significance of a moving total. If a moving total for sales drops it means the position is deteriorating, since such a drop indicates that the current period's sales fail to equal sales for the same period the previous year. A continuing fall indicates a continuing failure of current sales to equal the previous year's sales. Conversely, a rising total indicates an improvement. Of course, if costs are being considered, then a declining total indicates improvement in the form of reduced costs.

19. Graphing a MAT. It can be seen, therefore, that the use of moving totals helps to eliminate misleading impressions and if such totals are graphed, the slope of the curve will give a good indication of the immediate trend.

Two points, incidentally, should be noted in respect of graphing MATs:

(*a*) Since the object is to give a visual impression the graph is a presentational one. This means that the MATs will be plotted at the end of the year to which they relate, i.e. at the end of the last month that forms the annual total.

(*b*) Since the MAT will be some twelve times larger than the average month's figures, graphing actuals and MATs on the same scale can lead to difficulties. All too often the curve of the monthly figures will seem to creep insignificantly along the bottom of the graph. Possible solutions to this are to:

(*i*) use two vertical scales, one for the monthly figures and one for the MATs – care is needed in this case, however, since the two curves are not strictly comparable;

(*ii*) use a moving average (*see* next paragraph).

20. Moving average. A *moving average* is simply a moving total divided by the number of periods comprising the total. For example, look at Table 5B again. It shows the MAT throughout 19–6. Since each column is the sum of twelve months, the moving average is each moving total divided by 12, i.e.

Month	Mat		Moving Average
Dec (19–5)	73,200 ÷ 12	=	6,100
Jan (19–6)	72,200 ÷ 12	=	6,017
Feb	71,100 ÷ 12	=	5,925 etc.

21. Graphing moving averages. When graphing moving averages on a presentational graph then, since the object of using the moving average is simply to scale down the MATs, the averages can be plotted like MATs at the end of the year to which they relate.

Moving averages, however, are often used with mathematical graphs. In such a case the rule given in **14**(*b*)(*ii*) must be observed and so they must be plotted *at the midpoint of the period* to which they relate. Thus, the £6,100 just calculated above must be plotted at 30 June 19–5 and the £6,017 at 31 July 19–5.

In order to assist the correct plotting of moving averages it is a good idea to write the average opposite the midpoint of its period when constructing the table of moving averages. For instance, the January 19–6 figure of £6,017 (from **20**) will be written opposite the half-way point between July and August 19–5, thus:

Month	Actual	Moving average
July	6,200	
		6,017
August	6,400	

22. Advantages of moving values. The following are the advantages of moving values:

(*a*) they eliminate seasonal variations;

(*b*) when period figures fluctuate violently, moving values smooth out the fluctuations;

(*c*) if the values are moving averages then they can be plotted on the same graph as the period figures without a change of scale.

Z charts

Although graphs are generally designed for the particular purpose for which they are required, there are three types of graph so common that their forms have become standardised. They are Z charts, Lorenz curves and Band Charts, and these we look at in this and the next two sections.

23. Description of Z charts. A Z chart is simply a graph that extends over a single year and incorporates:

(a) individual monthly figures;
(b) monthly cumulative figures for the year;
(c) a moving annual total.

It takes its name from the fact that the three curves together tend to look like the letter Z.

24. Construction of a Z chart. Note that the following points apply in respect of the construction of a Z chart:

(a) Very often, as suggested in **19**(b)(i), a double scale is used on the vertical axis, one for the monthly figures and the other for the MAT and cumulative figures.

(b) As an example, Fig. 5.4 shows a Z chart for Fred's retail shop

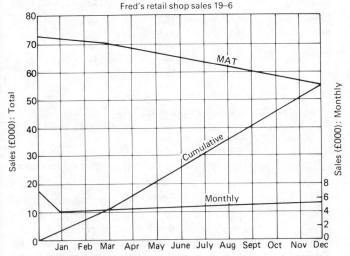

Figure 5.4 *Z chart*

sales figures (*see* Table 5A). Note where the different curves start:

(*i*) *monthly figures* at the December figure of the previous year;

(*ii*) *cumulative figures* at zero;

(*iii*) *MAT* at the MAT figure for the December of the previous year.

(*c*) Note that since a MAT is the total of the twelve immediately preceding months, the MAT for the final month must be the same as the cumulative total. The two curves will therefore meet at the last month of the graph.

Note: (1) The double vertical scales, one for the MAT and cumulative figures and the other for monthly figures.

(2) The figures needed to prepare the graph are as follows:

Sales (£000)

Month	Monthly figures	Cumulative total	MAT (see *Table 5B*)
Jan.	4.0	4.0	72.2
Feb.	4.1	8.1	71.1
Mar.	4.2	12.3	69.9
Apr.	4.3	16.6	68.6
May	4.4	21.0	67.2
Jun.	4.5	25.5	65.7
Jul.	4.6	30.1	64.1
Aug.	4.7	34.8	62.4
Sep.	4.8	39.6	60.6
Oct.	4.9	44.5	58.7
Nov.	5.0	49.5	56.7
Dec.	5.1	54.6	54.6

Note: (1) *Dec. 19–5 sales = 7.2 and MAT = 73.2.*

(2) The figures are far from being typical – *see* Progress test 5 question 4 for more typical figures.

Lorenz curves

It is a well-known fact that in practically every country a small proportion of the population owns a large proportion of the total wealth. Industrialists know too that a small proportion of all the factories employs a large proportion of the factory workers. This disparity of proportions is a common economic phenomenon, and a

Lorenz curve is a curve on a graph demonstrating this disparity.

To illustrate the procedure involved in the construction of a Lorenz curve, we will use the following data which typically shows that the largest number of retail sales are for very small amounts:

Table 5C *Customers' purchases in Fred's retail shop*

Value of customer's purchase	No. of purchases	Total purchase value (£)
Under 50p	310	105
50p to under £1	240	175
£1 to under £5	75	180
£5 to under £15	30	300
£15 and over	25	420

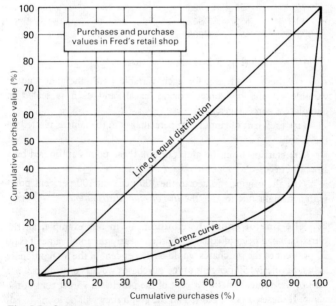

Fig. 5.5 *Lorenz curve*

25. Construction of a Lorenz curve. A Lorenz curve is constructed as follows (*see* Fig. 5.5).

(*a*) A table is drawn up from which the plots on the graph can be made. This table shows for each of the two series of variables involved:

(*i*) each variable value in the series;
(*ii*) each cumulative value in the series;
(*iii*) each cumulative value converted to a cumulative percentage.
The key to constructing Lorenz curves is remembering that it is the *cumulative percentages* that are to be plotted.

Table

Customer purchase			Purchase value		
No.	*Cum.*	*Cum.%*	*£*	*Cum.*	*Cum.%*
310	310	45.5	105	105	9.0
240	550	81.0	175	280	24.0
75	625	92.0	180	460	39.0
30	655	96.5	300	760	64.5
25	680	100.0	420	1180	100.0

(*b*) The graph (Fig. 5.5) is drawn so that:

(*i*) there is an axis for each variable (with the independent variable on the horizontal axis as usual) and each is scaled in cumulative percentages;
(*ii*) each pair of cumulative percentages in the table is plotted on the graph;
(*iii*) starting from the graph origin, these points are joined by a smooth curve;
(*iv*) the origin is joined to the 100 per cent/100 per cent point with a straight line to give the *line of equal distribution*.

26. The line of equal distribution. If, in our example, all the purchases had been of equal value then clearly the total value of (say) 25 per cent of the purchases would be 25 per cent of the total purchase value. Similarly, 50 per cent of the purchases would constitute 50 per cent of the value, and 75 per cent of the purchases 75 per cent of the value. If these pairs are plotted on the graph they will be found to fall on a straight line running from the origin to the 100 per cent/100 per cent point. Such a line, then, is *the curve which would be obtained if all purchases were of equal value*. It is called, therefore, the *line of equal distribution*.

27. Interpretation of Lorenz curves. The extent to which a Lorenz curve deviates from the 'line of equal distribution' indicates the degree of inequality. The further the curve swings away, the greater the inequality. There is no actual measure of this inequality but its extent can be gauged by reading the curve at the point where it lies furthest from the line of equal distribution and this in turn can be determined by observing the point at which the curve runs parallel to the line of equal distribution.

For example, in Fig. 5.5 the curve at its furthest point from the line of equal distribution is approximately at the 87 per cent cumulative 'Purchase' and 30 per cent cumulative 'Purchase value'. This means that 87 per cent of the purchases account for only 30 per cent of the total purchase value or, putting it the other way round, a mere 13 per cent of the purchases account for 70 per cent of the total purchase value.

28. The 80/20 rule. Lorenz curves of the pattern we have just seen are by no means uncommon. Since, where this pattern arises, the curve is typically at its maximum distance from the line of equal distribution at around the point 80 per cent on the horizontal axis and 20 per cent on the vertical axis, such occurrences follow the *80/20 rule* (Pareto phenomenon), which says that 80 per cent of one variable accounts for 20 per cent of the other. Thus, in Fred's retail shop, 80 per cent of the purchases accounted for approximately 20 per cent of the total purchases value. Similarly, it is often found that in an industrial store 80 per cent of the store items account for only 20 per cent of the total stores value, while nationally 80 per cent of the people own 20 per cent of the wealth (the other 20 per cent of the people owning 80 per cent of the wealth). In another context it is arguable that 80 per cent of trouble comes from 20 per cent of the trouble-makers.

29. Use of Lorenz curves. Lorenz curves can be used to show inequalities in matters such as:

  (*a*) incomes in the population;
  (*b*) tax payments of individuals in the population;
  (*c*) industrial efficiencies;
  (*d*) industrial outputs;
  (*e*) hospital patient costs;
  (*f*) customers and sales.

In some instances Lorenz curves can be used to compare *two series* of inequalities. For instance, if a second Lorenz curve, relating, say, to purchases in Fred's wholesale store, were superimposed on the curve for the purchases in Fig. 5.5, then it would be possible to compare inequalities of purchase values in these two enterprises and to see in which one purchase values were nearer to being equally distributed.

Band charts

Sometimes when the total figure in a time series has four or five component values, improved insight into the overall pattern can be obtained by showing the build-up of the total by a series of curves that divide the graph into bands. Such a graph is called a *band chart*.

To illustrate a band chart, assume that we have the following data relating to unemployment in Fredville:

Table 5D   *Fredville: no of people seeking employment*

			31 December:			
Time unemployed	*19–1*	*19–2*	*19–3*	*19–4*	*19–5*	*19–6*
Under 3 weeks	130	260	300	200	100	50
3 weeks – under 12 weeks	100	200	300	210	130	100
12 weeks – under 1 year	40	100	140	200	170	150
1 year and over	30	40	60	90	200	200
Total	300	600	800	700	600	500

30. Construction of a band chart. A band chart is constructed as follows (*see* Fig. 5.6).

(*a*) The order of the different bands is decided upon. In the case of the illustrative data the bands should logically show a time increase or decrease and the only question is whether the first band should be 'Under 3 weeks' or '1 year and over' (and to resolve this, assume we elect to start with the former). This is not always so. If, for instance, the data related to electricity generated from coal, nuclear fission, hydro-electric systems and solar power we would have to decide which source was to fill which of the four bands.

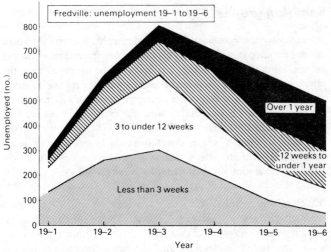

Figure 5.6 *Band chart*

(*b*) A table is prepared showing the cumulative totals at the top of each band. In the case of our data the table will be as follows:

Band	19–1 Act	Cum	19–2 Act	Cum	19–3 Act	Cum	19–4 Act	Cum	19–5 Act	Cum	19–6 Act	Cum
< 3 weeks	130	130	260	260	300	300	200	200	100	100	50	50
3–< 12	100	230	200	460	300	600	210	410	130	230	100	150
12–< 1yr	40	270	100	560	140	740	200	610	170	400	150	300
1 yr & over	30	300	40	600	60	800	90	700	200	600	200	500

(*c*) Each cumulative series is plotted as a separate curve on an ordinary graph.

(*d*) The bands are coloured or shaded so that they stand out distinctly. Note that as it is the *bands* which provide the information the curves in (*c*) should not be too thick.

(*e*) The bands are keyed – ideally within each band, otherwise at the side of the graph.

31. Reading a band chart. Reading a band chart is really no more than common sense, and so the reading of our illustrative chart is left for the reader to do in the next Progress test (*see* question 6).

Semi-log graphs

So far we have only considered linear scale graphs. Such graphs have a major limitation in so far as they give the wrong impression of *rates of change*. From the graph in Fig. 5.1, based on the figures given at the beginning of this chapter, it can be seen that sales of A have increased more than the sales of B. But the *percentage increase for the six years is the same*, namely 61 per cent (i.e. sales of A in 19–6 were 61 per cent greater than in 19–1, and so were sales of B).

If then we want a graph that shows which product's sales are increasing at the faster rate we cannot use a linear scale graph. To get the right impression we need a *semi-log graph* (or *chart*).

32. Semi-log graphs. A *log scale* is a scale on which *equal spans represent equal proportional changes*, instead of representing equal units. So, a span that represents change from 100 to 110 units (10 per cent) will, at the 1000 point on the scale, represent a change from 1000 to 1100 units. When one of the scales – normally the vertical scale – is a log scale while the other is a natural scale we say we have a *semi-log graph* (or *chart*).

33. Construction of a semi-log graph. A semi-log graph can be constructed in two ways:

(*a*) *By plotting the logarithm of a variable.* If the log of each vertical axis figure in the series is found and these *log* figures are then plotted on a linear scale graph, semi-log curves are obtained. In Fig. 5.7 figures from Fig. 5.1 and its table are again plotted on a semi-log graph. Note, incidentally, that although for plotting purposes the vertical axis must be log values, we can mark against the vertical scale equivalent linear scale values. While the object of a graph is to see the rate of change, this second scale does enable us to have some idea of the actual values represented by the curve.

(*b*) *By using semi-log paper.* Semi-log paper (which can be bought) is graph paper with the lines marking usual scale values drawn in logarithmic proportions (*see* Fig. 5.8 which is a graph with semi-log rulings). Such graph paper is recognisable by the changing distance between the horizontal lines. When using semi-log paper the vertical scale is marked with the *actual* values to be represented and the data plotted in the normal way – albeit care is needed reading the vertical scale (e.g. a point half way between two horizontal rulings has a value *less than* the mid point of the two values marked on the axis for the rulings).

34. Features of a semi-log graph. The following are the features of a semi-log graph (see Fig. 5.8):

(*a*) In the case of a single curve on a semi-log graph:

(*i*) the slope of the curve indicates the *rate* at which the figures are increasing (decreasing);

(*ii*) if the curve is a straight line, *the rate of increase (decrease) remains constant*, e.g. if the rate is 20 per cent per annum then the change will always be 20 per cent of the *previous* year's total – *see* curve *B*. A good example of such a curve would be one showing compound interest – although the amount of the interest is greater each year, it is always the *same percentage* of the total invested at the end of the previous period;

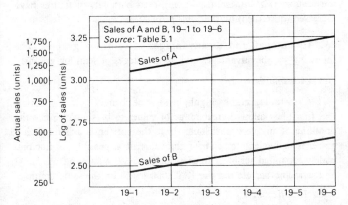

Figure 5.7 *Semi-log graph using linear scale.* The two curves are parallel, indicating that the rate of increase is the same for both, i.e. 10 per cent p.a.

(*iii*) if the *absolute* increase is constant, the curve will become progressively less steep (and progressively more steep if an absolute *decrease* is involved), e.g. if the increase is always (say) £5,000 per annum, the steepness must lessen, since £5,000 becomes a *continually smaller percentage of the increasing total figure* – *see* curve *A*.

The following are the figures from which the curves are drawn.

Year	Sales of A		Sales of B	
	Sales	Log	Sales	Log
19–1	1,121	3.0496	292	2.4654
19–2	1,233	3.0910	321	2.5065
19–3	1,356	3.1323	353	2.5478
19–4	1,492	3.1738	387	2.5877
19–5	1,641	3.2151	428	2.6314
19–6	1,805	3.2565	470	2.6721

(*b*) In the case of *two* curves plotted on a semi-log graph:

(*i*) the curve with the greatest slope has the greatest rate of increase or decrease – *see* curves *B* and *C*;

(*ii*) if they are parallel the rates of increase (decrease) are identical – *see* curves *D* and *C* in Fig. 5.8 and also 5.7;

(*iii*) if both curves over any part of the graph rise (fall) through the *same vertical distance* (say 1 cm), then both sets of figures have increased (decreased) by the *same percentage*.

35. Advantages and disadvantages of semi-log graphs. The following are the advantages and disadvantages of semi-log graphs:

(*a*) *Advantages:*

(*i*) Semi-log graphs highlight rates of change.

(*ii*) They allow a great range of values to be shown. Since the doubling of any vertical distance from the horizontal axis on such a graph is equivalent to *squaring* the value, it is possible to plot two widely separated series of figures – one series (say) around the 1,000 level and another around the 1,000,000 level – on the same graph.

(*b*) *Disadvantages:*

(*i*) If the reader of the graph is unfamiliar with the principle of semi-log graphs he may not interpret it correctly or even understand it.

(*ii*) It is impossible to show a zero on a semi-log graph. Series which include a zero (or negative figures) cannot be shown in full on these kind of graphs (and students should beware of trying to mark a zero on the vertical scale on a semi-log graph).

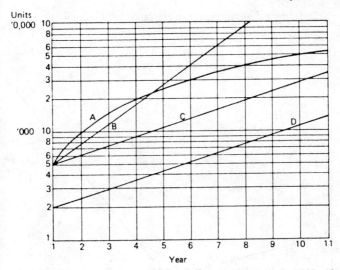

Figure 5.8 *Semi-log graph using semi-log rulings.* Curve A starts at 5,000 and increases by 5,000 every year. Curve B starts at 5,000 and increases by 50 per cent every year. Curve C starts at 5,000 and increases by 20 per cent every year. Curve D starts at 2,000 and increases by 20 per cent every year. Note that although Curve D is parallel to Curve C the actual size of each year's increase is considerably smaller, e.g. in year 2, 400 as against 1,000.

36. Log-log graph. Graphs are not confined to having the vertical scale alone a log scale. The horizontal axis can also be a log scale at the same time. Such a graph is, for obvious reasons, called a *log-log* graph. The use of these kinds of graphs, however, lies beyond the scope of this book.

Progress test 5

(*Answers in Appendix 4*)

1. The following figures relate to rainfall and the profits of an umbrella counter. Show both time series on the same graph:

Year	19-0	19-1	19-2	19-3	19-4	19-5	19-6	19-7	19-8	19-9
Rainfall (*cm*)	61	73	65	58	49	41	55	80	73	68
Shop profits (£)	621	740	894	773	702	591	488	661	992	863

2. Comment on the following graph:

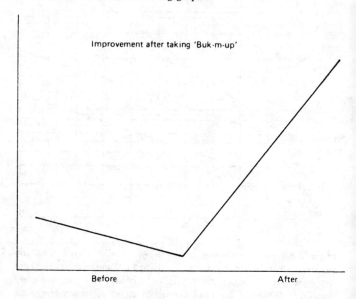

3. From the following annual figures calculate:

(*a*) the 3-year moving total;
(*b*) the 3-year moving average;
(*c*) the 10-year moving average.

Plot these, together with the individual annual figures, on the same graph. What is the difference between the 3-year moving average curve and the 10-year moving average curve?

Yearly figures 1960–84 (units)

1960 5	1965 8	1970 20	1975 9	1980 18
1961 8	1966 15	1971 16	1976 15	1981 22
1962 6	1967 10	1972 15	1977 8	1982 16
1963 12	1968 10	1973 6	1978 12	1983 14
1964 4	1969 13	1974 18	1979 14	1984 20

4. From the following data taken from monthly sales statistics construct a Z chart for 19–6 and comment on the graph:

Alpha Ltd sales (£)

Month	19–5	19–6	Month	19–5	19–6
Jan	15,000	17,000	July	6,000	11,000
Feb	14,000	19,000	Aug	6,000	1,000
Mar	11,000	18,000	Sept	8,000	5,000
Apr	10,000	18,000	Oct	10,000	5,000
May	8,000	18,000	Nov	10,000	8,000
June	7,000	12,000	Dec	13,000	10,000

5. The following figures come from a past Report on the Census of Production.

Textile machinery and
accessories

Establishments No.	Net output £000
48	1,406
42	2,263
38	3,699
21	2,836
26	3,152
16	5,032
23	20,385
214	38,773

Analyse this table by means of a Lorenz curve and explain what this curve shows.

(*CIMA*)

6. (*a*) What does the band chart in Fig. 5.6 tell you?

(*b*) What evidence is there that the figures have been contrived for this exercise?

Assignments

1. Plot the following time series on (*a*) a linear scale graph and (*b*) a semi-log graph (using, however, linear scale graph paper) and comment on the graphs.

Year sales	Group sales	Company A's
	(£000)	(£000)
19–2	1,620	135
19–3	1,780	154
19–4	1,950	176
19–5	2,140	195
19–6	2,350	208

2. Graph the following figures and superimpose the moving annual total:

Year	Qtr 1	Qtr 2	Qtr 3	Qtr 4
19–0	1,121	1,133	1,221	1,144
19–1	1,156	1,188	1,262	1,193
19–2	1,177	1,204	1,288	1,201
19–3	1,182	1,233	1,299	1,220
19–4	1,194	1,240	1,324	1,245
19–5	1,222	1,261	1,341	1,262
19–6	1,236	1,277	1,359	1,270
19–7	1,231	1,258	1,340	1,245
19–8	1,215	1,235	1,314	1,229
19–9	1,191	1,209	1,289	1,205

What conclusions can you draw from your graph?

6
Diagrams

Graphs, of course, are not the only way of presenting data visually. As we saw in 5.2 there are also diagrams which show data in essentially a one-dimensional way. The main forms of diagrams are as follows:

(a) *Pictorial representations.*

   (i) Pictograms.

   (ii) Statistical maps.

(b) *Bar charts.*

   (i) Simple bar charts.

   (ii) Component bar charts (actuals).

   (iii) Percentage component bar charts.

   (iv) Multiple bar charts.

(c) *Pie charts.*

Each of these will be considered in turn and illustrative examples will be taken from the data relating to the colours of the chairs sold in Fred's factory (*see* Table 4A).

Pictorial representation

1. Pictograms. This form of presentation involves the use of pictures to represent data. There are two kinds of pictogram:

(a) those in which the same picture, always the same size, is shown repeatedly – the value of a figure represented being indicated by *the number of pictures shown* (*see* Fig. 6.1 – note that a partly completed picture indicates a corresponding fraction of the block of units represented by a complete picture);

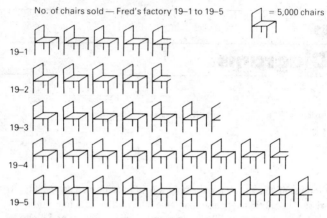

Figure 6.1 *Pictogram*

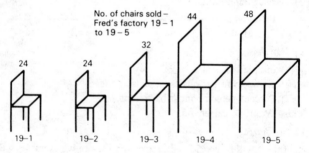

Figure 6.2 *Misleading pictogram*

(*b*) those in which the pictures change in size – the value of a figure represented being indicated by *the size of the picture shown* (*see* Fig. 6.2).

Type (*b*) is definitely not recommended as it can be very misleading. If the figure being represented doubles, for example, such an increase would probably be shown by doubling the height of the picture. However, if the height is doubled the width and length must also be doubled to keep the picture correctly proportioned and this results in the volume increasing by a factor of $2 \times 2 \times 2 = 8$. To the eye, then, it may well appear that the figure has increased by a factor of 8!

Sometimes an attempt is made to overcome the problem by simply doubling the volume, but inevitably there is always some confusion in the reader's mind whether heights or volumes, or even areas, represent the values depicted.

This type of pictogram, then, should only be used when exact proportions are of marginal interest only and the objective is more an artistic representation than a considered one. Note, however, that if this form is used the actual figures *must* be shown.

2. Statistical maps. These are simply maps shaded or marked in such a way as to convey statistical information (*see* Fig. 6.3).

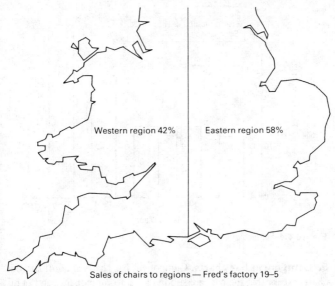

Western region 42%   Eastern region 58%

Sales of chairs to regions — Fred's factory 19–5

Figure 6.3 *Statistical map*

3. Uses of pictograms and statistical maps. These two types of diagram are very elementary forms of visual representation, but they can be more informative and more effective than other methods for presenting data to the general public who, by and large, lack the understanding and interest demanded by the less attractive forms of representation.

Bar charts

Bar charts are diagrams in which figures are represented by the lengths of the bars.

Since bar charts are similar to graphs, virtually the same principles of construction apply (*see* 5:**5–13**) – though note that here there should *never* be a 'break to zero' in bar charts.

4. Simple bar charts. In simple bar charts the data is represented by a series of bars, the height (or length) of each bar indicating the size of the figure represented (*see* Fig. 6.4).

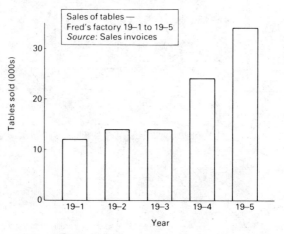

Figure 6.4 *Simple bar chart*

5. Component bar charts. Component bar charts are ordinary bar charts except that the bars are subdivided into component parts. This sort of chart is constructed when each total figure is built up from two or more component figures. They can be of two kinds:

(*a*) *Component bar chart (actuals).* In these charts the overall heights of the bars and the individual component lengths represent *actual* figures (*see* Fig. 6.5).

(*b*) *Percentage component bar chart.* In these charts the individual component lengths represent the *percentage* each component forms of

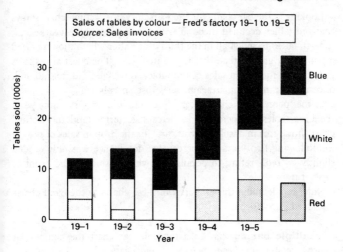

the overall bar total (*see* Fig. 6.6). Note that a series of such bars will all be the same total height, i.e. 100 per cent.

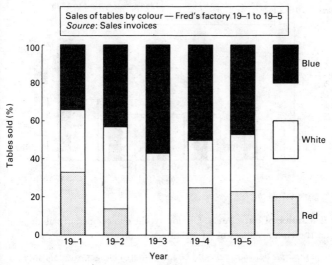

Figure 6.6 *Percentage component bar chart*

Looking at the bar charts for Fred's factory, the 'actual' chart shows that red tables declined in popularity for the first three years – and, indeed, failed to sell at all in the third year – after which they regained a popularity greater than that in the first year. The chart also shows that despite the fall in sales of red tables, total sales did not fall and, once red became popular again, sales rose sharply.

In the 'percentage' chart it can be seen that in the first three years the blue tables compensated in percentage terms (and the 'actual' chart shows this in absolute terms, too) for the drop in sales of red. In the following two years this percentage dominance was only eroded slightly by red's return to popularity – white absorbing the brunt of this return.

Note, incidentally, the similarity of these charts to the band charts discussed in 5:**30–31**.

6. **Multiple bar charts.** In a multiple bar chart the component figures are shown as *separate bars adjoining each other*. The height of each bar represents the actual value of the component figure (*see* Fig. 6.7).

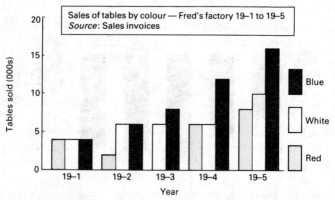

Figure 6.7 *Multiple bar chart*

7. **Choice of bar charts.** Obviously, the choice of chart will depend on the circumstances of its creation. Essentially:

(*a*) *simple bar charts* should be used where changes in totals only are required;

(b) *component bar charts (actuals)* should be used where changes in totals *and* an indication of the size of each component figure are required;

(c) *percentage component bar charts* should be used where changes in the *relative size only* of component figures are required;

(d) *multiple bar charts* should be used where changes in the actual values of the components figures *only* are required, and the overall total is of no particular importance.

Of course, for an all-round view there is no reason why all should not be adopted for, as we saw above, different forms bring out different features in the data.

Note that component and multiple bar charts can really only be used where there are not more than three or four components. More components make the charts too complicated to enable worthwhile visual impressions to be gained. Where a large number of components have to be shown, a pie chart (*see* **9** *et seq.*) is more suitable.

8. Bar charts v. pictograms. Bar charts are usually preferred to pictograms because:

(a) they are easier to construct;

(b) they can depict data more accurately;

(c) they can be used to indicate the size of *component figures*.

Pie charts

9. Pie chart definition. A *pie chart* is a circle divided by radial lines into sections (like slices of a pie, hence the name) so that the area of each section is proportional to the size of the figure represented (*see* Fig. 6.8). It is, therefore, a convenient way of showing the size of the component figures in proportion to each other and to the overall total.

10. Pie chart construction. Geometrically it can be proved that if the section angles at the centre of the 'pie' are in the same proportions as the figures to be illustrated, then the areas of the sections will also be in the same proportions. To construct a pie chart, then, it is only necessary to construct angles at the centre of the 'pie' in proportion to the figures concerned. Thus, taking the tables sold in 19–5, the 8,000 red tables constitute 23.5 per cent of the total number of tables and

Sales of tables — Fred's factory 19–5

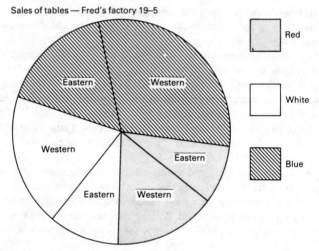

Figure 6.8 *Pie chart.*

hence an angle of 23.5 per cent of 360° = 84.5° (remember, the total number of degrees at the centre of the circle is 360). Similarly the 47 per cent blue tables section of the pie is drawn by constructing an angle of 47 per cent of 360°. And if we assume that the sales of the colours in the Western and Eastern regions were pro rata to the total sales in those regions, then the rest of Fig. 6.8 is quickly constructed.

11. Use of a pie chart. A pie chart is particularly useful where it is desired to show the relative proportions of the figures that go to make up a single overall total. Unlike bar charts, its effectiveness is not limited to three or four component figures but can extend up to seven or eight, thought it tends to diminish after that.

Pie charts, however, cannot be used effectively where a *time series* of figures is involved, as a number of different pie charts are not easy to compare. Again, note that changes in the overall totals should not be shown by changing the size of the 'pie', for the same reason as one should not change the size of pictograms.

Other diagrammatic forms

Many other diagrammatic forms can be used to present data visually.

One form employs maps drawn proportional to populations. The map outline is kept in its geographical shape as far as is possible but the areas of the various parts are made proportional to the populations residing in those parts. Figure 6.9 shows such a map depicting England, Scotland and Wales with Greater London superimposed on the England section. As can be seen, this particular map brings out the low populations of Scotland and Wales relative to England and especially Greater London, the population of which, in fact, is not much less than that of Scotland and Wales combined.

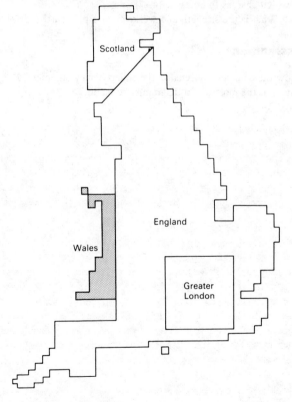

Figure 6.9 *Map of mainland Great Britain with countries and Greater London proportional in size to population.*

To conclude, it would appear that the only limitation to the various forms of diagrammatic presentation of data is that set by human ingenuity.

Progress test 6

1. What is the difference between a graph and a diagram? (5:**1**–**2**)

2. Distinguish between the correct and the incorrect forms of pictogram (**1**)

3. What kinds of bar charts are there, and what governs the choice of which chart is to be used? (**4**–**7**)

4. What is the strength and weakness of a pie chart? (**11**)

Assignment

Show the data you collected in the college data assignment, Progress test 4, in the most useful diagrammatic manner.

Part three
Frequency distributions

7
Preparing frequency distributions

Look at Table 7A. What information can be gleaned from this mass of figures?

Table 7A *Raw data*

Distances (km) recorded by 120 salesmen in the course of one week

482	502	466	408	486	440
470	447	413	451	410	430
469	438	452	459	455	473
423	436	412	403	493	436
471	498	450	421	482	440
442	474	407	448	444	485
505	515	500	462	460	476
472	454	451	438	457	446
453	453	508	475	418	465
450	447	477	436	464	453
415	511	430	457	490	447
433	416	419	460	428	434
420	443	456	432	425	497
459	449	439	509	483	502
424	421	413	441	458	438
444	445	435	468	430	442
455	452	479	481	468	435
462	478	463	498	494	489
495	407	462	432	424	451
426	433	474	431	471	488

Certainly it can be seen that most of the distances are in the 400s, though an occasional figure in the 500s is noticeable. But, once more, the limitations of the human mind make it impossible to detect whether there is any pattern in the figures. Are they spread out evenly, for example, or are there points of concentration? To answer such questions, statistical techniques can be used to process such a mass of figures relating to a single variable so that their significance can be grasped. This part of the book considers the various techniques involved.

Arrays and ungrouped frequency distributions

1. Raw data. The distances in Table 7A were obtained by simply listing the figures as they were given by the salesmen. Figures collected in this way are termed the *raw data* which can be defined as *data recorded as it is observed or received*. Note, incidentally, that as there is only a single characteristic being measured – distance – we have only one *variable* (*see* 1:**4**). For the rest of this part of the book, then, we will only be concerned with *single variable analyses*.

2. Array. The first obvious step to be taken in making the raw data more meaningful is to relist the figures in order of size, i.e. rearrange them so that they run from the lowest to the highest. Such a list of figures is called an *array*. An array of the distances in Table 7A is shown in Table 7B.

3. Ungrouped frequency distributions. An examination of the array in Table 7B suggests a further simplification. Since some figures repeat, e.g. 407, it would clearly simplify the list if each figure were listed once and the number of times it occurred written alongside, as in Table 7C.

In statistics the number of occurrences is called the *frequency* (and symbolised as f), and what we have in Table 7C is called an *ungrouped frequency distribution* ('ungrouped' simply distinguishes it from the grouped distribution to be discussed in the next section). An *ungrouped frequency distribution*, then, is a *list of the figures, in array form, occurring in the raw data, together with the frequency of each figure*.

Note that the sum of the frequencies (Σf) must equal the total number of items making up the raw data. (For the explanation of Σ *see* Appendix 1.)

Table 7B *Array of the raw data in Table 7A*

Distances (km) recorded by 120 salesmen in the course of one week

403	428	440	452	465	483
407	430	441	453	466	485
407	430	442	453	468	486
408	430	442	453	468	488
410	431	443	454	469	489
412	432	444	455	470	490
413	432	444	455	471	493
413	433	445	456	471	494
415	433	446	457	472	495
416	434	447	457	473	497
418	435	447	458	474	498
419	435	447	459	474	498
420	436	448	459	475	500
421	436	449	460	476	502
421	436	450	460	477	502
423	438	450	462	478	505
424	438	451	462	479	508
424	438	451	462	481	509
425	439	451	463	482	511
426	440	452	464	482	515

Grouped frequency distributions

While Table 7C (the ungrouped frequency distribution) is an
improvement on the array, there are still too many figures for the
mind to be able to grasp the information effectively. Consequently it
must be simplified even more. This can be done by *grouping* the
figures. When we group figures we obtain a grouped frequency
distribution (GFD). A grouped frequency distribution is the most
common and fundamental of all forms of presentation of statistical
data and so it is very important that all aspects of its construction are
well understood. More often than not, examination questions relating
to single variable analyses are presented in the form of a grouped
frequency distribution.

4. Grouped frequency distributions. One way of simplifying our

Table 7C *Ungrouped frequency distribution constructed from the array in Table 7B*

Distances (km) recorded by 120 salesmen in the course of one week							
Distance	Frequency	Distance	Frequency	Distance	Frequency	Distance	Frequency
403	1	434	1	456	1	479	1
407	2	435	2	457	2	481	1
408	1	436	3	458	1	482	2
410	1	438	3	459	2	483	1
412	1	439	1	460	2	485	1
413	2	440	2	462	3	486	1
415	1	441	1	463	1	488	1
416	1	442	2	464	1	489	1
418	1	443	1	465	1	490	1
419	1	444	2	466	1	493	1
420	1	445	1	468	2	494	1
421	2	446	1	469	1	495	1
423	1	447	3	470	1	497	1
424	2	448	1	471	2	498	2
425	1	449	1	472	1	500	1
426	1	450	2	473	1	502	2
428	1	451	3	474	2	505	1
430	3	452	2	475	1	508	1
431	1	453	3	476	1	509	1
432	2	454	1	477	1	511	1
433	2	455	2	478	1	515	1

Total frequency $(\Sigma f) = 120$

salesmen's data is to group the figures – and that involves showing how many frequencies occur in arbitrarily chosen bands or groups. We could, then, say that 12 frequencies were in the group 400 to under 420 km, and 27 in the group 420 to under 440 km. Such a group is called a *class* and a complete list of such classes together with their frequencies is called a *grouped frequency distribution*. Converting our ungrouped frequency distribution into a grouped frequency distribution gives us the figures in Table 7D.

Note that the choice of classes is in the hands of the statistician and an alternative form of grouping would be just as valid – though, as will be seen, care in selecting the groups has a considerable influence on

Table 7D *Grouped frequency distribution*

Distances (km) recorded by 120 salesmen in the course of one week	
Distance (km)	Frequency (*f*)
400 – under 420	12
420 – under 440	27
440 – under 460	34
460 – under 480	24
480 – under 500	15
500 – under 520	8
	120

the extent to which information about the distribution can be extracted.

5. Effect of grouping. As a result of grouping, it is possible to detect a pattern in the figures. For instance, the distances in Table 7D cluster around the '440–under 460' class. However, it is important to realise that, although it brings out the pattern, *such grouping results in the loss of information.* For example, the total frequency in the '400–under 420' class is known to be 12, but there is no longer any information as to *where in the class* these 12 occurrences lie. A clearer pattern has been bought at the cost of loss of information. The exchange is well worth while, but it means that calculations made from a grouped frequency distribution cannot be exact, and consequently excessive accuracy can only result in spurious accuracy.

6. Class limits. The *extreme boundaries of a class* are called *class limits.* Care has to be taken in defining class limits, otherwise there may be overlapping of classes or gaps between classes. Imprecision here is a common fault. Given, say, classes of 400–420 and 420–440 km, in which of the two would a distance of 420 km be recorded? Obviously, it could be either. Conversely, if we were told that the higher class were 421–440, a distance of 420.5 km would appear to fit into neither of them. This kind of imprecision is very unsatisfactory – grouping in itself results in enough loss of information without ambiguity losing more. To avoid this, classes are frequently stated so that the top limit of one class is all but the bottom limit of the next class. So the first class in Table 7D is '400–under 420 km', i.e. the class limits are exactly 400 km at the lower end and *right up to, but not including* 420 km (the lower limit of the next class) at the upper end, i.e. 419.9999. A well-designed frequency distribution will always ensure that there is neither overlapping of classes nor gaps between them.

7. Stated and true class limits. The *class limits actually given in a grouped frequency distribution* are called the *stated limits*. The stated limits are not necessarily the true limits. For example, if a distribution were headed 'Age last birthday' and the first class were 10–19 years, i.e. stated limits of 10 years and 19 years, we would know from common sense that this meant the *true* limits were 10 years exactly and 19 years, 364 days. (This must be since anyone who had passed his 19th birthday but who had not reached his 20th birthday was 19 last birthday and so would be recorded in the 10–19 years class.) In this sort of situation it is very important for students to remember that one should *always use the true limits and never the stated ones* – and that the true limits are found by using common sense.

Formally, *true limits* can be defined as the *true underlying limits of the distribution classes in respect of the real world from which the data was drawn*. There will, of course, be many occasions when the stated limits will also be the true limits but one should always check first before using the stated limits in any statistical analysis.

8. Discrete and continuous data. Next the reader should be warned that data exists in one of two forms, discrete or continuous.

Discrete data is data that increases *in jumps*. For instance, if the data relates to the numbers of children in families then the figures recorded will be 0, 1, 2, 3, or 4, etc. 1.5 or 2.25 children are impossible figures. In other words, fractions of a unit are impossible and the data increases in jumps – from 0 to 1, from 1 to 2, etc. (Note, however, the units themselves can be fractions, e.g. 0.5p.)

Continuous data on the other hand is data that can increase *continuously*. If, say, kilometres travelled are being investigated, the figures recorded could end in any fraction of a kilometre imaginable, e.g. 425.001, 425.634, 425.999, etc.

Now it is very important to note that whether data is discrete or continuous *depends solely upon the real nature of the data and not upon how it is collected.* Thus, the distances travelled by the salesmen discussed so far are continuous data since any fraction of a kilometre can occur in reality, although a glance at the raw data (Table 7A) reveals no fractions at all, i.e. they were recorded as if they were discrete data. To repeat, it is not how the data is recorded that matters but only how it occurs in reality, and on this basis alone one decides if data is continuous or discrete.

9. Mathematical limits. Let us next consider the true class limits of a distribution of discrete data. Say, for example, the *stated* limits of a class are 5–under 10. Clearly, if the data is discrete there can be no occurrences above 9 and below 10, and so the *true* limits of the class are 5–9. It also follows that the true lower limit of the next class is 10.

Now according to the rule given in **7** we should use these true limits in our statistical calculations. However, for purely mathematical reasons it is very awkward working with a gap between classes – albeit in nature there really is a gap. Consequently we close the gap by extending the true limits of each class by half a unit – i.e. our '5–9' class now becomes 4.5–9.5 and our next class starts at 9.5. These extended limits are called *mathematical limits,* and *if our data is discrete we must always work with the mathematical limits.* (Note, incidentally, that a class with true limits 0 to 4 will have mathematical limits of −0.5 to 4.5. Do not let this −0.5 bother you. It is purely a mathematical limit and you can be assured that at the end of any computations involving this negative limit everything will come out as it should do.)

Note that for reference all rules relating to class limits are diagrammatically summarised in Fig. 7.1.

10. Class limits and the form of raw data. In **7** we said that in determining the true class limits attention should be paid to the reality underlying the raw data. For instance, in the raw data of salesmen's distances shown in Table 7A there are no fractions of a kilometre. Yet it is inconceivable that the distances travelled by so many salesmen could all be exact kilometres. There must have been some rounding.

Now, if the figures were rounded to the nearest kilometre, 419.75 kilometres, for example, would be recorded as 420 kilometres. That means it would be grouped in the '420–under 440' class in Table 7D. *But clearly such a distance should be in the '400–under 420' class.*

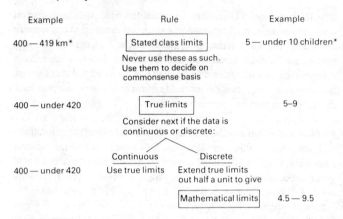

Example	Rule	Example
400 — 419 km*	Stated class limits Never use these as such. Use them to decide on commonsense basis	5 — under 10 children*
400 — under 420	True limits Consider next if the data is continuous or discrete:	5–9
400 — under 420	Continuous — Discrete Use true limits — Extend true limits out half a unit to give Mathematical limits	4.5 — 9.5

* Note that the stated class limits have deliberately been stated in a misleading way.

Fig. 7.1 *Class limits rules.*

Therefore, if the figures have been rounded to the nearest unit Table 7D would be incorrectly constructed. Its construction can only be correct if the raw data were recorded on the basis of the number of *completed* kilometres.

This illustration shows how important it is to reflect carefully on the way the raw data is collected. The errors in subsequent analyses which could result from incorrectly constructing a grouped frequency distribution would in most cases be small – particularly in view of the error which is inevitable following the loss of information on grouping. However, there is no justification in compounding an acceptable error with an unnecessary error and so care should be taken in the construction and interpretation of any grouped frequency distribution.

11. Class interval. A *class interval* is the *width of the class concerned*, i.e. the difference between the class limits (the true or mathematical class limits, of course, depending upon whether the data is discrete or continuous). If the class intervals of the classes are equal, the distribution is said to be an *equal class interval distribution*. In Table 7D the class interval is 20 kms for all classes and the distribution is therefore an equal class interval distribution.

12. Unequal class intervals. Some sets of figures are such that, if equal class intervals were taken, a very few classes would contain nearly all the occurrences while the majority would be virtually empty, e.g. a distribution of annual incomes with class intervals of £10,000. In cases like this it is better to use unequal class intervals. These should be chosen so that the overfull classes are subdivided and the near-empty ones grouped together, e.g. £0–under £4,000, £4,000–under £5,000, £5,000–under £6,000, £6,000–under £8,000, £8,000–under £10,000, £10,000–under £15,000, £15,000–under £25,000, £25,000–under £50,000, £50,000 and over. Fairly obviously, such a distribution is called an *unequal class interval distribution*.

13. Open-ended classes. If the first class in a distribution is stated simply as 'Under . . .', e.g. 'Under 400 kilometres' or the last is stated as 'Over . . .', e.g. 'Over 500 kilometres', such classes are termed *open-ended*, i.e. one end is open and goes on indefinitely. They are used to collect together the few extreme items the values of which extend way beyond the main body of the distribution.

The *class interval of an open-ended class* is by convention deemed to be the same as that of the class immediately adjoining it. In well-designed distributions, open-ended classes have very low frequencies and so the error that arises from using the convention is not that important, although a careful selection of the class limits at the upper and lower ends of the distribution can ensure that even this error is kept to an absolute minimum.

14. Choice of classes. It goes without saying that the construction of a grouped frequency distribution always involves making a careful decision as to what classes will be used. The choice will depend on individual circumstances, but the following points should be borne in mind.

(*a*) Classes should be between ten and twenty in number.

Note: For the sake of simplicity the number of classes in examples in this book will be kept very small.

(*b*) Class intervals should be equal wherever practicable.

(*c*) Class intervals of 5, 10 or multiples of 10 are more convenient than other intervals such as 7 or 11.

(*d*) Classes should be chosen so that occurrences within the classes tend to balance around the midpoints of the classes. It would be unwise to have a class of (say) £8,000–under £8,100 in a salary distribution, since salaries at this level are often in round £100s, so most of the occurrences would be concentrated at £8,000, i.e. the extreme end of the class. This is unsatisfactory, since later theory makes the assumption that the average of the occurrences in a class lies at the midpoint of the class.

15. Direct construction of a grouped frequency distribution. Once the raw data has been recorded (as in Table 7A) we may wish to construct a grouped frequency distribution directly, without going through the intermediate steps of an array and an ungrouped frequency distribution. To make such a direct construction the following steps should be taken:

(*a*) Pick out the highest and lowest figures (in Table 7A these are 403 and 515) and on the basis of these and a general overview of the whole of the raw data decide upon and list the classes.

(*b*) Observe each figure in the raw data and insert a tally mark (|) against the class into which it falls (*see* Table 7E). Note that every fifth tally mark is scored diagonally across the previous four. This simplifies the totalling at the end.

(*c*) Total the tally marks to find the frequency of each class.

The distribution is now complete, though the class frequencies should be added up and the total checked to see that it corresponds with the total number of items in the raw data.

Table 7E *Direct construction of Table 7D from raw data* (see *Table 7A*)

Class	Tally marks	Frequencies
400–under 420	|||| |||| ||	12
420–under 440	|||| |||| |||| |||| |||| ||	27
440–under 460	|||| |||| |||| |||| |||| |||| ||||	34
460–under 480	|||| |||| |||| |||| ||||	24
480–under 500	|||| |||| ||||	15
500–under 520	|||| |||	8
		120

Progress test 7

(*Answers in Appendix 4*)

1. Suggest classes for insertion into the *Classes* column of grouped frequency distributions compiled from raw data relating to:

   (*a*) a survey of the ages of adults in a city – questionnaires were sent to all people of 20 years and over, asking them to state their present age in years;

   (*b*) a survey of the number of extractions made on a specific day by a group of dentists, the numbers ranging between 4 and 32;

   (*c*) incomes per annum of all full-time employed adults in a town (recorded to the nearest £).

   State the exact class limits of the second class chosen by you for each of the distributions.

2. (*a*) Reconstruct the grouped frequency distribution for Table 7A using the same class interval but starting with the first class at 390 kilometres.

   (*b*) If the data in this example had been rounded to the nearest kilometre as suggested in **10**, what would be the true limits of the class '430–under 450 km'?

Assignment

Visit your nearest library, select a bay of shelving and observe the number of pages in 200–300 books. Set out this data in the form of a frequency distribution.

8
Graphing frequency distributions

Although a grouped frequency distribution enables us to see some pattern in the figures a visual representation again affords us an even better grasp of the pattern. In this chapter the various ways of visually representing a grouped frequency distribution are explained.

Histograms

First we look at histograms and right from the start it must be clearly understood that although histograms often look like bar charts they are *not* bar charts. Where there are bar charts it is *heights* that are the crucial measure but where there are histograms it is areas that are crucial.

1. **Construction of a histogram**. A *histogram* is the graph of a frequency distribution. It is constructed on the basis of the following principles:

(*a*) The horizontal axis is a continuous scale running from one extreme end of the distribution to the other. This means that this axis is *exactly the same as any ordinary axis on a graph*. It should be labelled with the name of the variable and the units of measurement.

(*b*) For each class in the distribution a vertical rectangle is drawn with:

(*i*) its base on the horizontal axis extending from one class limit of the class to the other class limit;

(*ii*) its *area proportional to the frequency in the class*, i.e. if one class has a frequency twice that of another, then its rectangle will be twice the area of the other.

Note: (1) The class limits at the ends of the rectangle base are the true limits in the case of continuous data and the mathematical limits in the case of discrete data. (This means, incidentally, that there will *never be any gaps* between the histogram rectangles.)

(2) If the distribution is an *equal class interval distribution* (the more usual case) then the bases of all the rectangles will be the same widths. This means that to obtain areas proportional to frequencies the heights must be drawn proportional to the frequencies, i.e. a class having twice the frequency of another will have a rectangle twice the height, and in this particular case (*but in this case only*) the vertical axis of the graph can be labelled 'Frequency'.

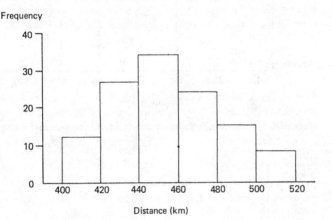

Figure 8.1 *Histogram of Table 7D data.*

The histogram for Table 7D is shown in Fig. 8.1. As this is an equal class interval distribution the heights of the rectangles are in this case drawn proportional to the frequencies of the classes.

2. Distributions with unequal class intervals. If, in Table 7D, classes 400–under 420 and 420–under 440 were merged to give a single class 400–under 440, the combined frequency would be 39.

Now suppose we construct a histogram and that the rectangle for this group is drawn two class intervals wide with a height of 39. The resulting graph (Fig. 8.2) is seriously in error. Compared with the original histogram (Fig. 8.1), it is obvious that the area of the combined classes is much greater than the combined areas of the two separate classes.

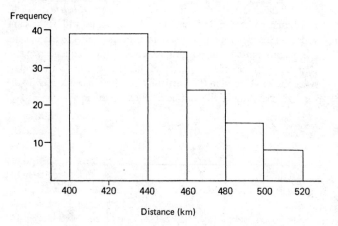

Figure 8.2 *Histogram with unequal class intervals – incorrect construction.*

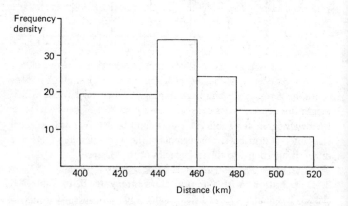

Figure 8.3 *Histogram with unequal class intervals – correct construction.*

As it is essential that the total areas should be the same, an adjustment must be made. This adjustment is quite simple, for since the 400–under 440 class has a class interval double that of the other normal classes, its histogram rectangle has a width double that of the other rectangles. Now, area = width × height, so a doubling of the width can be adjusted for by halving the height. Thus the correct histogram is drawn with a frequency of 19.5, i.e. half of 39, for the 400–under 440 class (*see* Fig. 8.3).

Note: Compared with the original histogram, this average height still does not look quite correct. This distortion arises through the loss of information as a result of further grouping.

3. Frequency density. The observant reader may have noticed that in Fig. 8.3 the vertical axis has changed its label from 'Frequency' to 'Frequency density'. That this axis no longer measures frequency obviously follows from the fact that although the first rectangle measures a height of 19.5, the total frequency of the class 400–under 440 is actually 39. However, the reason for the new axis label may not be so clear. It is because the heights of the rectangles now measure the *frequency density*, i.e. they measure how densely packed the occurrences are within each class, e.g. although class 400–under 440 has 39 occurrences these are spread over a 40 km class interval and hence are 'packed' together at an average density of 19.5 per 20 km, while the class 440–under 460 has 34 occurrences spread over a 20 km interval, i.e. on average 34 occurrences per 20 km.

We can say, then, that the rectangle heights in a histogram really measure the *intensity of clustering*. This means that in the case of an unequal class interval distribution a unit of density (an area equal to a frequency of 1) must be selected, the vertical axis given a scale reflecting this density and each rectangle drawn on its base with a height that represents its density, i.e. with an area proportional to its frequency.

Frequency polygons and frequency curves

A histogram is a graph of rectangles. Instead of rectangles it may be decided to show a single curve rising and falling.

4. Frequency polygon. A curve for use in lieu of a histogram can be drawn in the following way:

(*a*) Add a class at each end of the grouped frequency distribution having a class interval equal to the class interval of its adjoining class and with a frequency of zero (which, of course, is the actual frequency of these classes). The reason for this apparently unnecessary increase in the number of classes is essentially mathematical – it is important that the curve reaches the horizontal axis at each end, and including these two zero frequency classes achieves this.

(*b*) Construct the histogram of the extended grouped frequency distribution ('drawing' rectangles of zero height for the two end classes).

(*c*) Mark the midpoints of the tops of each rectangle.

(*d*) Join the midpoints with straight lines.

The resultant curve gives us a figure known as a *frequency polygon*. Figure 8.4 is in fact a frequency polygon of Table 7D superimposed on our earlier histogram.

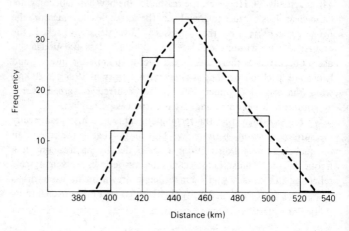

Figure 8.4 *Frequency polygon of Table 7D distribution.*

5. Area of a frequency polygon. It will be appreciated that in drawing a frequency polygon the corner of each rectangle cut off by the polygon is exactly equal in area to the triangle added between the point where the polygon line emerges from one rectangle and the midpoint of the next. Thus, the area of the frequency polygon is exactly equal to the area of the histogram.

6. Frequency curve. If the frequency polygon is smoothed so that there are no sharp points, it is known as a *frequency curve*. Normally, frequency curves should only be constructed when there are a large number of classes and very small intervals – when, in fact, the frequency polygon is almost a smooth curve anyway. A frequency curve constructed under other circumstances tends to be inaccurate.

Again, it should be appreciated that the area under the frequency curve must be equal to the area of the histogram on which it is based. The reason for this emphasis on areas, incidentally, is that in later statistical theory the areas of distributions are of crucial importance.

Finally, note in passing that since the vertical axis measures frequency density the peak of a frequency curve will indicate the point of maximum frequency density or clustering.

Ogives

Next, we look at ogives.

7. Ogive. The name *ogive* is given to the curve obtained when the *cumulative* frequencies of a distribution are graphed. It is also called a *cumulative frequency curve*.

8. Ogive construction. To construct an ogive:

(*a*) compute the cumulative frequencies of the distribution, i.e. add up the progressive total of frequencies class by class (*see* Fig. 8.5, which illustrates an ogive constructed from the data in Table 7D);

(*b*) prepare a graph with the horizontal axis as before and with the cumulative frequency on the vertical axis;

(*c*) plot a starting point at zero on the vertical scale and the lower class limit of the first class;

(*d*) plot the cumulative frequencies on the graph at the *upper class limits* of the classes to which they refer;

Note: This latter point is very important and must be clearly understood. Consider what a cumulative frequency figure means. In

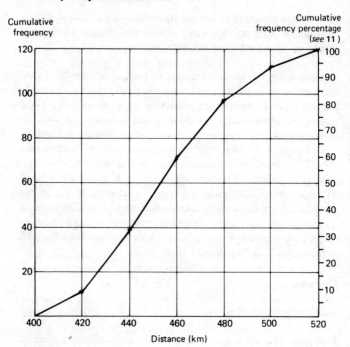

Figure 8.5 *Ogive*. Ogive of Table 7D data ('less than' curve). The table below shows the cumulative frequency distribution of the data in Table 7D.

Class	Frequency	Cumulative frequency	Cumulative percentage (see 11)
400–under 420	12	12	10
420–under 440	27	39	32.5
440–under 460	34	73	61
460–under 480	24	97	81
480–under 500	15	112	93
500–under 520	8	120	100
	120		

Fig. 8.5, for example, the cumulative frequency alongside class 420–under 440 is 39. This means that there is a total of 39 distances which lie below 440 kilometres. Therefore the 39 must be plotted just below the 440 kilometre point (for practical graphing purposes, at 440). Similarly, the 73 relating to the 440–under 460 class means a total of 73 distances below 460 kilometres and the 73 must therefore be plotted just below the 460 kilometre point. In other words – to repeat – when constructing an ogive, each cumulative frequency must be plotted at the upper class limit of its class.

(*e*) join all the points.

9. 'Less than' curves. Take any value on the cumulative frequency scale – say 100. If you now read off the distance that the curve gives for this cumulative frequency you will find it is about 484km. This means that 100 frequencies are less than 484km, i.e. 100 salesmen travelled distances less than 484km. Now take any value on the horizontal axis – say 430km. Reading the corresponding cumulative frequency value from the curve you obtain about 25. And this means that 430km was travelled by less than 25 salesmen.

In view of the way the curve is read it is hardly surprising that it is called a 'less than' curve. If desired, a 'more than' curve can be constructed by cumulating the frequencies in the reverse direction, i.e. starting with the frequency '8' of the class 500–under 520 (so at the end the curve will run from 120 at 400km to zero at 520km).

10. Estimating individual item values. Since we can use the 'less than' curve to say that as 100 of the total distances travelled are less than 484 kilometres, then clearly the 100th distance must be very, very close to 484 kilometres. Indeed, since estimates made from a grouped frequency distribution can never be completely accurate (*see* 7:**5**) we can say that to all intents and purposes the 100th distance figure *is* 484 kilometres.

This means, then, that we can use the ogive to estimate the value of any given item in the distribution. All we need to do is to identify the item on the cumulative frequency axis, run our eye across to the curve and then down to the horizontal axis where the value of that item can be read off, e.g. the value of the 20th item in our distribution can be seen to be about 425 kilometres – and, conversely, 500km was the 112th distance travelled.

11. Comparing ogives. It is sometimes desired to compare the ogives of two different distributions. Unless the total frequencies of the distributions are the same, such comparisons are virtually impossible if the ogives are constructed as outlined above. One ogive will simply tower over the other. The difficulty is easily overcome, however, by plotting *not* the cumulative frequencies themselves but the cumulative frequencies *expressed as a percentage of the total frequencies in the distribution*. This results in both ogives ending at the same point, 100 per cent, and so renders them comparable (*see* Fig. 8.5, noting both the percentage column in the table and the percentage scale on the right-hand side of the graph).

12. Smoothing ogives. The ogives discussed so far have consisted of a series of straight lines. However, in practice the curve of a cumulative frequency distribution tends to change its slope smoothly and not in sharp kinks. So, providing that it passes through all the plotted points, it is better to draw a smooth curve than an angular one. The reason that in practice the curve slope changes smoothly is that occurrences are not usually spread equally throughout a class but tend to cluster towards adjoining classes that have higher frequency densities. Smoothing automatically makes some allowance for this tendency in the case of ogives and therefore improves the accuracy of the curve.

Progress test 8

(*Answers in Appendix 4*)

1. Construct (*a*) a histogram, (*b*) a smooth ogive ('less than' curve) for the following data:

In a certain examination, 12 candidates obtained fewer than 10 marks, 25 obtained 10 to under 25 marks, 51 obtained 25 to under 40 marks, 48 obtained 40 to under 50 marks, 46 obtained 50 to under 60 marks, 54 obtained 60 to under 80 marks, and only 8 obtained 80 marks or more. Marks were out of 100.

2. From Fig. 8.5 determine: (*a*) the number of distances recorded below 470 kilometres; (*b*) the range of distances that were recorded by the lowest 25 per cent of the salesmen.

Assignments

1. Construct histograms for the following data and superimpose frequency polygons on them.

Marks scored in IQ tests by pupils at two different schools

IQ marks	Number of pupils	
	School A	School B
75–under 85	15	43
85–under 95	25	99
95–under 105	40	54
105–under 115	108	40
115–under 125	92	14
125 and over	20	0
	300	250

2. Prepare a histogram and frequency polygon for the data in the assignment in Chapter 7.

3. Prepare a histogram and frequency polygon for the distributions in Appendix 7.

9
Fractiles

In this chapter we see how further knowledge of the distribution under analysis can be gained by ascertaining values of individual selected items.

Before going further, however, an important point must be made. In this and the next two chapters a whole variety of measures will be discussed – fractiles, means, standard deviations, etc. It is important to appreciate that all the values which are computed will be *in the same units as the distribution from which they are found*. So if this is in kilometres then the measure will be in kilometres, if in £s, then in £s, and if in numbers of monkeys, then in numbers of monkeys.

Fractiles of ungrouped frequency distributions

Although, normally, fractiles are computed from grouped frequency distributions, the concept behind this kind of measure is better grasped if ungrouped frequency distributions are first looked at. This we do in this section using the array in Table 7B (where each of the six columns contains 20 items) to illustrate the different fractiles. Note, incidentally, that the term 'distribution' implies an array form of the data.

1. Fractiles. A *fractile* is the *value of the item which is a given way through a distribution*.

If you are asked to give the value of an item one-third of the way through an array of data, all you would do would be to start at the beginning of the array and count until you came to the item one-third of the way through it. You would then simply give the value of that item as your answer. And this value is the required fractile.

Here one point needs great emphasis, and that is that a fractile is the *value* of the fractile item. *It is not the fractile item.* Thus, in our illustrative distribution of 120 distances, the required fractile would be the *value* of the 40th item – it would not be the 40th item itself.

2. Fractiles and ogives. When we examined ogives we found they could be used for estimating the value of any given item (*see* 8:**10**). Clearly then, ogives can be very useful for estimating fractiles – all we need to do is to find which item is the fractile item and use the ogive to read off its value immediately.

Example

Find the value of the item one-sixth of the way through our illustrative distribution in Table 7D. This fractile is the value of the 120/6 = 20th item, and reading directly from the ogive in Fig. 8.5 the value of this item can be seen to be about 425 kilometres.

Although using an ogive in this way is an excellent method of estimating fractiles, an alternative method is by calculation. This alternative method is discussed in **11–14** so as to provide the student with two different ways of finding fractiles.

Both, incidentally, give equally correct answers – unless the examination question calls for one or the other the student may select either. In practice, calculation generally proves quicker if finding only one or two fractiles, while the ogive is quicker if more than two fractiles are required.

3. Median. Obviously, one can take any fraction of a distribution one wants – third, quarter, three-tenths, etc. – and find the corresponding fractile, but the most common fraction is a half. The fractile obtained using half as the required fraction is called the *median*, which can therefore be defined as the *value of the middle item of a distribution*. Thus, in a distribution of 120 items the median is the *value* of the 60th item.

To find the median then:

(*a*) arrange the data in an array;

(*b*) compute which item is the median (middle) item by dividing the total distribution frequency by 2 – rounding *up* if the answer contains a half (e.g. the middle item of 9 is the $9/2 = 4\frac{1}{2}$th, and rounded up this gives the 5th item);

(c) read off from the array the value of this median item.

In Table 7B distribution the 60th item is 452km.

4. Median item in an even-numbered distribution. Strictly speaking, the 60th item referred to above is not the middle item in a distribution of 120 items. There is, in fact, no middle item, the midpoint of the distribution lying between the 60th and the 61st items. (If this is not quite clear, imagine a distribution with only four items. Which is the middle item? Clearly, there is not one, the midpoint of the distribution lying between items 2 and 3.)

The fact that an even-numbered distribution has no single midpoint item often bothers students but in practice there is no difficulty, since in any realistic distribution the values of the middle two items will be virtually the same – as in the case of our illustrative distribution – and either will suffice (if they are not then it would be unwise to use the median at all). Alternatively, the values of the middle two items may be averaged.

5. Use of the median. The median is a useful statistical measure for as it is the value of the item in the middle of a distribution it means that, generally speaking, half the items have values above the median and half have values below the median. Since a few items may have the same value as the median (particularly if the data is discrete) it is more accurate to say that half the items have a value *equal to or greater than* the median and half a value *equal to or less than* the median.

6. Quartiles. Quartiles are the fractiles relating to the one-quarter and three-quarters fractions. The *quartiles*, then, can be defined as the values of the items one-quarter and three-quarters of the way through a distribution. The value of the item one-quarter of the way through the distribution is called the *lower* or *first quartile* (symbolised as Q_1) and the item three-quarters of the way the *upper* or *third quartile* (symbolised as Q_3).

In the case of our salesmen the first quartile item is item $120/4 = 30$th, and it can be seen that this has a value of 434km. The third quartile item is $3 \times 120/4 = 90$th which has a value of 473km.

7. Use of quartiles. The median is a valuable statistical measure in its own right, but because it is the value of only one item out of a distribution of possibly hundreds, it often needs supplementing by other measures. One of its shortcomings, for instance, is that it gives no indication as to how far on either side of it the other values extend.

To know that the median wage in a department is £116 per week gives us no idea of the actual level of pay received by most of the lower-paid employees, except that it is not above £116. If, however, it is also known that the first quartile is £105 and the third £130, then it is clear that many employees earn a wage close to the median, while some earn much more (a quarter of the employees must earn £130 or more by the very definition of a quartile). Indeed, it is possible to go further. Since a quarter of the distribution lies below the first quartile and a quarter lies above the third, then *half the distribution must lie between the two quartiles*. It can be concluded, then, that half the employees earn between £105 and £130 per week, whilst a quarter earn £105 or less and a quarter earn £130 or more.

From this it can be seen that a considerable amount of information will be contained in a statement that gives only three figures, if these are the median and the two quartiles. For this reason, if it is wished to summarise a large set of figures very briefly, the median and the two quartiles may be chosen as the representative figures that carry the most information.

8. Deciles. *Deciles* are fractiles relating to *10ths* of the way through a distribution. Thus, the first decile is the value of the item one-tenth of the way through a distribution (in the case of our distribution, the 120/10 = 12th item, which has a value of 419km). The second decile is the value of the item two-tenths of the way (so the $120 \times 2/10 = 24$th item, which has a value of 430km) and the, say, ninth decile is the item nine-tenths of the way, i.e. the 108th item in our distribution, and this has a value of 494km.

9. Percentiles. *Percentiles* are fractiles relating to *100ths* of the way through a distribution. Thus the 85th percentile is the value of the item 85/100ths of the way through a distribution. So in our distribution the 85th percentile item is item $120 \times 85/100 = 102$nd and so the 85th percentile is 485km.

10. Use of percentiles. Percentiles are particularly useful as 'cut-off' values. For example, assume that only 22 per cent of boys applying for a place in school can be accepted and that acceptance is based on a particular test. If the boys' results in this test are set out in an array, the 78th (100−22) percentile will indicate the test score that will separate the successful applicants from the others.

Fractiles of grouped frequency distributions

As was pointed out in 7:5, grouping data results in a loss of some information since once the data is grouped it is no longer possible to know what the value of an actual item is. Since fractiles are all about values of actual items this creates a minor problem. However, if item values are intelligently estimated acceptable answers can be found.

In this section we look at fractiles in relation to grouped frequency distributions and for illustrative purposes we will now use the grouped frequency distribution shown in Table 7D.

11. Computing the fractile items. The computation of the fractile *item* remains unchanged from its computation in the case of an ungrouped frequency distribution. So, if we want the first quartile of a grouped frequency distribution of 1000 items, the first quartile item is the 1000/4=250th item.

12. Estimating the value of a fractile item. Let us assume we wish to estimate the value of the 60th item of our illustrative distribution, i.e. the fractile item is the 60th. To do this we argue as follows:

(*a*) Although we don't know exactly the value of the 60th item, if we prepare a cumulative frequency table we can identify in just which class our fractile item falls. And the cumulative frequency distribution of our illustrative distribution is shown in Table 9A.

So, if we start counting our frequencies from the lower end of our distribution we will find that by the time we are just under 440 we will have counted 39 items and by the time we are just under 460 we will have counted 73 items. This means the 60th item must lie in the class 440–under 460. This class we call the *fractile class*.

Table 9A *Cumulative frequency distribution of Table 7D data*

Kilometres	Frequency	Cumulative frequency
400–under 420	12	12
420–under 440	27	39
440–under 460	34	73 (fractile class)
460–under 480	24	97
480–under 500	15	112
500–under 520	8	120

(*b*) Next, if 39 items have been counted by the time 440km has been reached, then the 60th item lies 60−39 = 21 items further on (and from now on the argument is also illustrated diagrammatically in Figure 9.1). As Table 9A shows, there is a total of 34 items in this class. So, if the items are spread equally throughout the class (and we have to make this assumption if we are to make any progress), our fractile item lies 21/34ths of the way into that class.

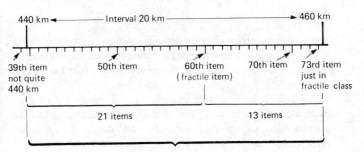

Figure 9.1 *Locating the fractile item within the fractile class.*

(*c*) Continuing, we see from Table 9A that the class interval of our fractile class is 20km (460−440). So our fractile item lies 21/34ths through this interval of 20km—in other words, it lies 20 × 21/34 = 12.4km (approximately) into the interval.

(*d*) Finally, since the lower limit of the fractile class is 440km, our best estimate of the *value* of our fractile item is 440 + 12.4 = 452.4km.

So our fractile is 452.4km.

13. Procedure for estimating fractiles from a grouped frequency distribution. From the above arguments we can summarise here the procedure and formula for estimating any desired fractile from a grouped frequency distribution as follows:

(*a*) Find the *fractile item* by multiplying the desired fraction by the total number of items in the distribution and then rounding up (e.g. the third quartile item in a distribution of 110 items = ¾ × 110 = 82.5, which is rounded up to give the *83rd item*).

(*b*) Prepare a cumulative frequency distribution from the grouped frequency distribution and locate the *fractile class*, i.e. the class containing the required fractile.

(*c*) Apply the following formula:

Required fractile = Lower class limit of fractile class +

$$\left(\frac{\text{Fractile item} - \text{cumulative frequency up to}}{\text{Fractile class frequency}} \times \begin{matrix}\text{Fractile} \\ \text{class} \\ \text{interval}\end{matrix} \right)$$

Three points should be noted in estimating fractiles in this manner.

(1) Clearly a correct estimate depends upon correctly determining the lower class limit of the fractile class. As usual, the true limits must be used for continuous data and the mathematical limits for discrete data.

(2) When a fractile is estimated from a grouped frequency distribution of discrete data the answer frequently contains a fraction of a unit. Since the object of estimating a fractile is to find an actual single value – that of the fractile item – and since such a fraction is impossible in the case of discrete data, it is necessary to round the answer to the nearest unit. If, for example, the data in Table 9A had related to numbers of cars, the median value of 452.4 would have been rounded to 452 cars.

(3) The procedure is as equally applicable to unequal class interval distributions as equal class interval distributions.

14. Estimating medians, quartiles, deciles and percentiles. Since medians, quartiles, deciles and percentiles are all fractiles the procedure above can be used to estimate each of these measures from a grouped frequency distribution.

To illustrate the procedure in these instances we will apply it to the estimate of the lower and upper quartiles, the 9th decile and the 85th percentile of our illustrative distribution, Table 7D (as the median item is the 120/2-60th item, it follows that in **12** we were, in fact, finding the median of the distribution).

Table 9B *Estimating fractiles from a grouped frequency distribution*

			1st Quart.	3rd Quart.	9th Decile	85th %ile
Fractile item:			120/4 = 30	$3 \times \dfrac{120}{4} = 90$	$9 \times \dfrac{120}{10} = 108$	$85 \times \dfrac{120}{100} = 102$
Class	*f*	*Cum f*		*Fractile class*		
400–<420	12	12				
420–<440	27	39	420–440			
440–<460	34	73				
460–<480	24	97		460–480		
480–<500	15	112			480–500	480–500
500–<520	8	120				
No. of items into fractile class			30–12 = 18	90–73 = 17	108–97 = 11	102–97 = 5
Fractile class frequency			27	24	15	15
Fractile class interval (km)			20	20	20	20
∴ Km into fractile class			20 × 18/27 = 13.3	20 × 17/24 = 14.2	20 × 11/15 = 14.7	20 × 5/15 = 6.7
Lower limit, fractile class			420	460	480	480
∴ Fractile (km)			433.3	474.2	494.7	486.7

As a matter of interest these values (together with the median value of 452.4km) may be compared with those found in paragraphs **3–9** to illustrate the error which follows as a result of the loss of information on grouping.

Progress test 9

(*Answers in Appendix 4*)

1. The following marks were obtained by candidates in an examination (no fractions of a mark were awarded):

Marks	No. of Candidates	Marks	No. of Candidates	Marks	No. of Candidates
0–5	2	36–40	150	71–75	120
6–10	8	41–45	200	76–80	100
11–15	20	46–50	220	81–85	60
16–20	30	51–55	280	86–90	40
21–25	50	56–60	320	91–95	17
26–30	80	61–65	260	96–100	3
31–35	120	66–70	160		
					2240

(a) Find: (i) the median, (ii) the first quartile, (iii) the third quartile, (iv) the sixth decile, (v) the 42nd percentile. (*Warning:* think carefully about class limits.)

(b) If the examining body wished to pass only one-third of the candidates, what should the pass mark be?

2. The distribution of gross annual earnings for 200 operatives in an engineering company is shown below.

Range of earnings (£)		Number of operatives
At least	Less than	
	3,000	7
3,000	4,000	30
4,000	5,000	37
5,000	6,000	51
6,000	7,000	32
7,000	8,000	25
8,000	10,000	14
Over 10,000		4
Total		200

You are required to:

(a) draw a cumulative ('less than') frequency distribution on squared paper;

(b) use your diagram to find estimates of the median, upper quartile, lower quartile and highest decile.

(*CIMA*)

Assignments

1. Find the median, two quartiles, third decile and 33rd percentile of the data assignment in Chapter 7. Comment on these measures.

2. Find the medians and upper and lower quartiles of the distributions in Appendix 7.

10
Averages

Assume someone has prepared a distribution about which we know nothing – the lifespan, say, of a newly discovered type of microbe. We do not want to study the distribution in detail but simply want a few single figures that will give us a good idea of what the distribution is like. What figures would be useful?

Well, first we have no idea as to whereabouts on a timescale, running from zero to eternity, the microbe lifespan distribution lies. A figure giving us some information as to the *location* of the distribution would clearly be useful. Next, it would be helpful to know if the distribution was spread out over a large part of the timescale, or whether in fact the lifespans were clustered closely together. In other words, a single figure summarising the *dispersion* of the distribution would be useful. Finally, we may like to know if the majority of the lifespans lie at one end of the distribution or the other, or whether they are ranged symmetrically around the middle of the distribution, i.e. we would like a figure indicating the *skew* of the distribution.

In this chapter we will examine measures of location of a distribution and in the next chapter measures of dispersion and skew.

Medians, means and modes

1. Measuring location – the average. Finding a single figure that indicates the location of a distribution is not easy. The distribution may be spread out over a wide range of values, and to choose one to represent its location is like trying to find a single person who can represent a whole constituency in Parliament. Just as there are different opinions as to which person would make the best constituency representative, so there are different opinions as to which figure best represents the location of the distribution. Note that any representative of such a location is called an *average* or, more formally, a *measure of central tendency*.

In statistics there are three main averages, each in a different way measuring the location of the distribution. These are the *median*, the *arithmetic mean* (henceforth simply referred to as the 'mean') and the *mode*.

2. Median. We have already come across the median (*see* 9:**3**). Since the median was defined as the value of the middle item of distribution it has an obvious claim to be an average.

The median has the following features that make it a particularly useful statistical measure.

(*a*) As we have seen, half the items in a distribution have a value equal to or above the median and half equal to or below the median. (This in turn means that if we choose an item from the distribution at random there would be a 50/50 chance that its value would be equal to or above the median, and vice versa.)

(*b*) It is unaffected by the value of extreme items in the distribution. Given what was said in 3:**10** it will be appreciated that any statistical measure that does not rely on the correctness of extreme items for its value possesses a useful advantage.

3. Finding the median. Finding the median was discussed in 9:**3** and 9:**12**.

4. Mean. The *mean* (or the 'arithmetic mean', to give it its full title) is the measure to which we usually refer in everyday life when we use the word 'average', and it can be defined as the value each item in the distribution would have if all the values were shared out equally among all the items. Thus, if three people had £2, £3 and £7 respectively the mean amount would be £4, i.e. £12 shared equally between three people. The greatest advantage the mean possesses is that in calculating it every value in the distribution is used and this means that it can be used for further statistical computations, e.g. averaging means (*see* **17**).

5. Finding the mean. The arithmetic mean (symbolised as $\bar{x}$ and referred to as 'bar x') can be found from the formula $\bar{x} = \Sigma x/n$, where x stands for the values of the different items in the distribution.

If we use this formula as it stands we are said to be using the *direct method* of computing the mean. For example, if we have the figures 5, 7, 8, 12, 18, then using the direct method we have $\bar{x} = \Sigma x/n = (5+7+8+12+18)/5 = 10$.

In the case of a grouped frequency distribution, this method obviously cannot be used since the values of the individual items are not known. Instead we need to adopt the following procedure:

(*a*) lay out the grouped frequency distribution;
(*b*) find the midpoint of each class;
(*c*) multiply the class midpoint by the class frequency (remember, the class frequencies should ideally lie equally either side of the class midpoint – and so we can assume that collectively the class frequencies all approximately equal the value of the midpoint of their classes);
(*d*) add these products and divide by the total distribution frequency.

Example
Find the mean of the distribution of salesmen's distances shown in Table 7D.

Kms	f	Class midpoint (MP)	f×MP
400–under 420	12	410	4,920
420–under 440	27	430	11,610
440–under 460	34	450	15,300
460–under 480	24	470	11,280
480–under 500	15	490	7,350
500–under 520	8	510	4,080
	120		54,540

$\therefore \quad \bar{x} = 54,540/120 = 454.5\text{km}.$

6. Interpreting the mean. By and large the mean is a very misunderstood measure. Often it is regarded as being much the same as the median, so if the average (mean) wage is declared to be £200 per week, many people earning substantially less are puzzled. Yet if nine people earn £100 per week and one person earns £1,100 then the mean wage for the ten people is $(9 \times 100 + 1100)/10 = £200$, and this despite the fact that 90 per cent of the workers earn only half the 'average'. Another apparent paradox was illustrated some years ago when an Australian test bowler, who always batted last, never scored more than seven or eight runs each time he batted, but ended the season with an average of over 100 runs an innings. The fact was that on all but a single occasion the batsman at the other end was always out first, so the bowler's average was his total runs divided by one, i.e. over 100.

Given this capability to mislead it is fair to ask just what does the mean figure tell us? And the answer is that it tells us how much or how many we would have in total if we were told that there were a given number of items having a specified mean value. In other words, the total value is $n\bar{x}$ – i.e. if 10 people have an average wage of £200 a week then between them their wages total $10 \times £200 = £2,000$. And it should be fully appreciated that it is unwise to read much more into the mean than this.

7. Mode. Finally, we come to what is computationally the least satisfactory of the three main averages – the mode. The *mode* can be defined as the *point of maximum frequency density*. As we saw in 8:**3**, the vertical axis of a histogram strictly speaking measures frequency density. So where one has a frequency curve the point of maximum frequency density occurs under the peak of the curve, and this is how we identify the mode. Three aspects of this definition, however, immediately arise – first, how to proceed where there is more than one peak; second, how to find the peak when only a simple histogram exists; and third, how to find the mode when a frequency curve has no meaning.

We look at these in reverse order.

8. Mode of a discrete limited distribution. Assume we have a distribution of the number of children in nuclear families in Western Europe. In such a distribution the variable is both discrete and limited, i.e. it would not extend at the upper end to much above ten. In this situation you cannot have a frequency curve since there cannot be fractional values (even though one could be drawn, readings from it would have no significance). So here the point of maximum frequency density is quite simply the discrete value having the highest rectangle in the histogram and – as a moment's thought will show – this will be the most frequently occurring value. In these circumstances, therefore, the *mode* is defined as *the most frequently occurring value in a distribution.*

9. Identifying the mode from a histogram. Where there is a distribution of continuous data, or discrete data with so many steps as to be virtually continuous, the peak of its frequency curve can be estimated in the following way (*see* Fig. 10.1 where the mode of our illustrative grouped frequency distribution is identified):

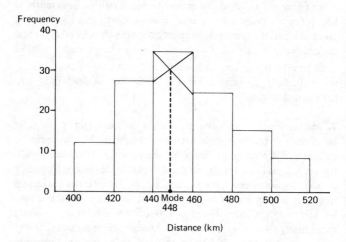

Figure 10.1 *Diagrammatic method of finding the mode.* The histogram uses the data in Table 7D.

(*a*) Construct the histogram of the grouped frequency distribution.
(*b*) Identify the class with the highest rectangle. This is called the *modal class* since it is the class that contains the mode.
(*c*) Draw a line from the top right-hand corner of the modal class rectangle to the point where the top of the next adjacent rectangle to the left touches it. Draw a corresponding line on the opposite diagonal from the top left-hand corner to the rectangle on the right.

(*d*) Read off on the horizontal axis where the two lines cross. This is the mode.

Example
Find the mode of our illustrative data.
See Fig. 10.1 which shows that the mode is 448km.

10. Bi-modal distributions. Should the frequency curve of a distribution be such that two peaks arise, the distribution is said to be *bi-modal*. Distributions normally have a single mode, but a distribution relating to (say) wage rates in a factory that employs both men and women could be bi-modal, one mode relating to the rate paid to the majority of women and the other to that paid to the majority of men. In such a case, then, there is no single figure for the mode (do not, under any circumstances, average the two modes), and if asked for the mode you should give both as your answer.

As a matter of interest bi-modal distributions often occur when, unknown to the compiler, the distribution is really two separate distributions combined as in the wages example just given. If, unexpectedly, you should find that you have a bi-modal distribution you should, therefore, look carefully to see if you cannot disentangle two quite distinct subdistributions.

11. Computing the mode. Where there is a single frequency curve peak for continuous, or virtually continuous, data then another way the mode can be estimated is from the following formula:

Mode = Mean − 3(Mean−Median)

Note that if the expression inside the brackets is positive, the mode lies below the mean, whereas if it is negative it lies above the mean.

Example
Mode of our illustrative distribution = $454.5 - 3(454.5 - 452) = 454.5 - 7.5 = 447$km

An alternative formula is:

Mode = Lower limit$_m$ + Class interval$_m (f_m - f_{m-1})/((1 f_m - (f_{m-1} + f_{m+1}))$
where m stands for the modal class and $m-1$ and $m+1$ the classes above and below m.

Example
$$\text{Mode} = 440 + 20(34 - 27)/(2 \times 34 - (27 + 24)) = 440 + 140/17$$
$$= 448.2 \text{km}$$

The difference between our two answers indicates how precise a mode derived from a computation is in practice.

12. Validity of a point modal estimate. As was implied in the previous paragraph, on the whole, apart from the case of discrete limited data, a point modal estimate often has doubtful validity. If it is found graphically from a histogram then it is quite possible that if the distribution were to be rewritten from the original data using different classes, a different mode will emerge. And the validity of the computation does depend on the distribution being fairly typical (and not, of course, bi-modal – a feature not always observable if only the mean and median are known).

This being so, then unless one is sure that the mode is a valid figure it is better in practice simply to state the *modal class*, i.e. the class with the highest histogram rectangle.

13. Interpretation of a mode. Being a peak of a frequency curve, or better, the most frequently occurring value in a discrete distribution, the mode is the best representative of the *typical* item. It is this form of average that is implied by such expressions as 'the average person' or 'the average holiday'. The remark that 'the average holiday is four weeks' means that the *usual* holiday is four weeks, not that the mean of all holidays is four weeks. The mode is thus a familiar and commonly used average, though its name is less well known.

As we have seen, where the data is continuous the modal class is a better representation of the typical than a point modal estimate. It is much more reasonable to say, for instance, that the typical wage packet in an enterprise is between £130 and £140 a week than to say it is £132.52.

Mean, median and mode compared

Now that we have seen how the three different averages are found we can see how they compare.

14. Choice of average. It is sometimes difficult to know which average should be used in any given situation. Essentially:

(*a*) if we wish to know the result that would follow from an equal distribution of the values – consumption of beer per head, for instance – the *mean* is the most suitable;

(*b*) if the half-way value is required, with as many above as below, the *median* will be the choice;

(*c*) if the most typical value is required then the *mode* is the appropriate average to use.

To illustrate the use of the different averages we can consider their applicability to a distribution of, say, doctors' salaries. If, first, we are interested in the purchasing power of a group of doctors then we need to know the *mean* since any normal group of doctors will collectively earn the mean salary multiplied by the number in the group. On the other hand, if we are discussing medicine as a career with a particular school-leaver, then the *median* will be the most useful average since we can say that, unless our school-leaver is special in any way, he will have a 50/50 chance of obtaining the median salary or better. Finally, if we are an organisation providing a service to doctors and charging on the basis of so much per doctor we should bear the *mode* in mind as this is the salary of the typical doctor and the acceptance or rejection of our service may well depend on our charge in relationship to the doctor's salary.

15. Features of the mean, median and mode. The features of our three averages can be summarised as follows:

(*a*) *The mean.*

(*i*) The mean is the best known of the averages.

(*ii*) It makes use of every value in the distribution. It can, therefore, be distorted by extreme values.

(*iii*) It can be used for further mathematical processing (e.g. *see* **17**).

(*iv*) It may result in an 'impossible' figure where the data is discrete, e.g. 1.737 children.

(*b*) *The median*

(*i*) The median is equal to or exceeded by half the values in the distribution – and vice versa.

(*ii*) It uses only one value in the distribution. It is *not*, therefore, influenced by extreme values.

(*iii*) It cannot be used for further mathematical processing.

(*iv*) It is an actual value occurring in the distribution (unless it is computed by averaging the two middle items of an even-numbered distribution).

(*v*) It can be computed even if the data is incomplete. Thus, in determining the median salary of a group of executives, for example, it may prove impossible to discover the salaries of the highest-paid executives. But, since these values will not affect the value of the median item, it is still possible to determine the median salary.

(*c*) *The mode.*

(*i*) If the data is discrete then, like the median, the mode is an actual, single value.

(*ii*) If the data is continuous then the mode marks the point of the greatest clustering of occurrences.

(*iii*) Like the median, it can often be computed from incomplete data.

(*iv*) Again like the median, it cannot be used for further mathematical processing.

Averaging means

Sometimes we want to find the average (mean) of two or three means. For instance, if 10 students in one class have a mean of 60 marks and 40 students in another class have a mean of 50 marks, what is the average mark of the two classes combined?

16. How not to average means. In this sort of problem there is a strong temptation to compute the average of the means, i.e. $\frac{1}{2}(60+50)=55$ marks. *This is quite wrong* (although if the groups all comprise an identical number of items then the answer, although computed improperly, will in fact be correct).

17. Procedure for averaging means. In establishing a procedure for averaging means the following argument is used:

(*a*) (*i*) The total mark gained by the whole of the first class must be 60×10, i.e. 600 marks.

Note: Since $\overline{x} = \Sigma x/n$, therefore $\Sigma x = n\overline{x}$, where Σx is the total of all the values in the distribution.

(*ii*) Similarly, the total mark gained by the second class must be $50 \times 40 = 2{,}000$ marks.

(*b*) The combined marks of both classes, then, must be $600 + 2{,}000 = 2{,}600$ marks.

(*c*) These marks were obtained by a total of $10 + 40 = 50$ students, so the mean mark per student must be $2{,}600/50 = 52$ marks.

So the *correct* procedure for computing such a combined average is as follows:

(*a*) Compute for each group the sum of all the values. This is done by multiplying the group mean by the number of items in the group.

(*b*) Add these totals to give a combined total.

(*c*) Divide the combined total by the combined number of items.

18. Weighted average. This average mean is often referred to as the *weighted average* because individual group averages are 'weighted' by multiplying them by the number of items in the group.

It should be noted that this particular use of a weighted average is really only a specific application of a more general idea. There are other contexts in which, when computing a mean, it is sometimes desired to give certain figures greater importance in the answer, e.g. index numbers. This can be done by multiplying those figures by chosen numbers called *weights*. In the example above, the weights were the number of students in each class, i.e. 10 and 40. The important point to remember about a mean computed from such weights is that, instead of dividing by the number of items, one divides by *the sum of the weights* (unweighted figures carry a 'weight' of 1 in this addition). A mean so calculated is called a *weighted average*, and the appropiate formula is:

$$\text{Weighted average} = \frac{\Sigma xw}{\Sigma w}$$

Example

Find the weighted average of the following figures where Group B figures are to carry weights of 2, and Group C figures, weights of 3:

Group A: 6, 5, 3
     B: 12,14
     C: 20, 22, 23

Group	No.(x)	Weight (w)	Weight × No. (xw)
A	6	1	6
A	5	1	5
A	3	1	3
B	12	2	24
B	14	2	28
C	20	3	60
C	22	3	66
C	23	3	69
		$\Sigma w = 16$	$\Sigma xw = 261$

Weighted average $= \dfrac{\Sigma xw}{\Sigma w} = \dfrac{261}{16} \simeq 16.3$

19. Accuracy and the mean. Since the mean is only a representative value of a possibly large number of very different values, it would seem pedantic to state it with extreme accuracy, even though such accuracy may not be spurious. For instance, to give the mean height of a group of children as 1.24m conveys just as much to a reader as 1.2402m. However, in statistics the mean is frequently used in further calculations (as above, where it was used to obtain a weighted average). For this reason, it is necessary to state it with a good deal more accuracy than would otherwise be called for.

Geometric and harmonic means

There are two other means apart from the arithmetic means which, though not used anything like as often as the other averages, sometimes arise in examinations. These are the geometric mean and the harmonic mean.

20. Geometric mean (GM).

  (*a*) *Computation in theory:*

    (*i*) multiply the values all together, i.e. $x_1 \times x_2 \times x_3 \ldots$, and then, (*ii*) find the *n*th root of the product where *n* is the number of items.

Since many modern calculators enable almost any kind of root to be found, the geometric mean can often be computed directly using the above procedure. However, if such a calculator is not available then there is a simple alternative – though longer – method of computation.

(*b*) *Calculator-free computation.* Providing log tables are at hand the computation of the geometric mean becomes one of simple arithmetic since with logs multiplication is achieved by adding and an *n*th root is found by dividing by *n*. So the formula becomes:

$$\text{Logarithm of geometric mean} = \frac{\Sigma \log x}{n}$$

To compute the GM therefore, we must:

(*i*) add the logs of all the values;
(*ii*) divide by *n*;
(*iii*) look up the anti-log of the answer to (*ii*).

Example
Find the geometric mean of 4.5 and 6.
With calculator: $\text{GM} = \sqrt[3]{4 \times 5 \times 6} = 4.933$
Without calculator: $\text{Log GM} = \dfrac{\log 4 + \log 5 + \log 6}{3}$

$$= \frac{0.6021 + 0.6990 + 0.7782}{3} = \frac{2.0793}{3} = 0.6931$$

$\therefore$ GM = anti-log $0.6931 = 4.933$

(*c*) *Use of the GM.* The geometric mean is used mainly in connection with index numbers (*see* 15:**24**), although it can also be used for averaging ratios.

21. Harmonic mean.

(*a*) *Computation.*
(*i*) add the reciprocals of the values; then
(*ii*) divide this sum into the number of items.

The formula is:

Harmonic mean $= n/\Sigma \dfrac{1}{x}$

Example

Find the harmonic mean of 4, 5 and 6.

Harmonic mean $= \dfrac{3}{\frac{1}{4} + \frac{1}{5} + \frac{1}{6}} = \dfrac{3}{37/60}$

$= 4.86$

(*b*) *Use of the harmonic mean.* The use of the harmonic mean in statistics is so restricted that discussion is best omitted in a handbook such as this. However, if an examination question should ask for the average of different *speeds* then the harmonic mean should be computed.

Example

A car travels from A to B at an average speed of 60 mph and returns at an average speed of 40 mph. What is its average speed for the entire journey?

Harmonic mean $= n/\Sigma \dfrac{1}{x}$

$= 2/ \left(\dfrac{1}{60} + \dfrac{1}{40} \right) = 48$ mph

Progress test 10

(*Answers in Appendix 4*)

1. In a series of 20 'spot checks' the following number of passengers were counted at a certain depot:

```
137  136  135  136
135  135  137  138
136  137  136  136
138  137  136  137
136  136  138  135
```

(*a*) From these figures determine: (*i*) the mean, (*ii*) the mode, and (*iii*) the median.

(*b*) It is later discovered that the last observation was incorrectly recorded when the data was being collected. It should have been 35 instead of 135. Recompute the three averages. What features of the three averages do your revised figures bring out?

2. Using the data in Assignment 1 in Chapter 8.

(*a*) Compute the mean and median IQ of the pupils in each school.
(*b*) If 450 pupils in a third school had a mean IQ of 106, what would be the mean IQ of all the pupils in the three schools combined?

3. During the 1977–78 session a college ran 70 different classes of which 44 were 'science', with a mean class size of 15.2, and 26 were 'arts', with a mean class size of 19.2. The frequency distributions of class size were as follows:

Size of class (number of students)	Number of science classes	Number of arts classes
1– 6	4	0
7–12	15	3
13–18	11	10
19–24	8	8
25–30	5	4
31–36	1	1
	44	26

No student belonged to more than one class.

Required:
(*a*) Calculate the mean class size of the college.
(*b*) Suppose now that no class of 12 students or less had been allowed to run. Calculate what the mean class size for the college would have been if the students in such classes:

(*i*) had been transferred to the other classes;
(*ii*) had not been admitted to the college.

(*c*) The number of students enrolling in 1980–81 on science and arts courses is expected to rise by 20 per cent and to fall by 10 per cent respectively, compared with 1977–78. Calculate the maximum number of classes the college should run if the mean class size is to be not less than 20.

(*ACCA*)

4. Customers' waiting times in a checkout queue at Fred's self-service store were found to be as follows:

Length of wait (min.)	No. of customers
No waiting	50
Waiting under $\frac{1}{2}$	210
$\frac{1}{2}$ – under 1	340
1 – under 2	200
2 – under 3	110
3 – under 5	170
5 – under 10	140
10 and over	80
	1,300

(*a*) A few lucky customers have no waiting and a few unlucky ones have to wait over 10 minutes. How long do the middle 50 per cent of the customers have to wait?

(*b*) (*i*) If you use the store every day, what will your mean waiting time be?

(*ii*) You wish to reach the checkout point by 11.15 a.m. It is now 11.13 as you pick up your last purchase. Are the odds in your favour?

5. A company which makes and sells a standard article has four machines on which this article can be made. Owing to differences in age and design the machines run at different speeds, as follows:

Machine	Number of minutes required to produce one article
A	2
B	3
C	5
D	6

(*a*) When all machines are running, what is the total number of articles produced per hour?

(*b*) Over a period of 3 hours, only machines B, C and D were run for the first 2 hours, and only machines A, B and D for the last hour. What was the average number of articles produced *per hour* over this 3-hour period?

(CIMA)

6. A distribution of the wages paid to foremen would show that, although a few reach very high levels, most foremen are at the lower levels of the distribution. The same applies, of course, to most wage distributions.

(a) If you were an employer resisting a foreman's wage claim, which average would suit your case best?

(b) If you were the foremen's union representative, which average would you then select?

(c) If you were contemplating a career as a foreman, which average would you examine?

Give reasons for your answers.

7. Show by means of a formula the weighted average of three groups: the first group having n_1 items and a mean of $\bar{x}_1$; the second group, n_2 items and a mean of $\bar{x}_2$; and the third group, n_3 items and a mean of $\bar{x}_3$.

Assignments

1. Find the mean and mode for the data in the assignment in Chapter 7.

2. Find the means and modes for the distributions in Appendix 7. Comment on these measures.

11

Dispersion and skew

Finally in this part of the book we look at two further features which relate to frequency distributions and can be measured – dispersion and skew.

Dispersion

Knowing the average of a distribution in no way tells us whether or not the figures in the distribution are clustered closely together or well spread out.

For example, there could be two groups, each of four men. In the first group, all the men could be 1.676m high, and in the second group could have heights of 1.372m, 1.524m, 1.827m and 1.981m respectively. Both groups have a *mean* height of 1.676m, but the dispersion in height is much greater in one than the other.

It would clearly be useful to find some way of measuring this dispersion and expressing it as a single figure. Such measures are called *measures of dispersion* (or *variation*) and the most important of these are the:

(a) range;
(b) quartile deviation (semi-interquartile range);
(c) mean deviation;
(d) standard deviation.

Note: The word *variation* is sometimes used as an alternative to dispersion.

1. Range. The *range* is simply the *difference between the highest and the lowest values*. Therefore:

Range = Highest value − Lowest value

Example

Find the range of the data in Table 7B.

Range = 515 − 403 = 112 kilometres

Unfortunately the range has a grave disadvantage: it is too much influenced by extreme values. If, for example, one single salesman in the Table 7B data had travelled 627 kilometres, the range would have been doubled although the dispersion of the other 119 salesmen would have remained unaltered. Since an extreme value often arises as a result of an error in the first place (*see* 3:**10**), the range can not only be a distorted measure but also one that is totally divorced from the reality of the situation.

2. Quartile deviation. The disadvantage of the range in respect of extreme values can be overcome by ignoring the extreme values. One way of doing this is to cut off the top and bottom quarters by considering only the quartiles, and then see what range is left. This range is called the *interquartile range*. If the interquartile range is divided by two, the figure obtained is called the *quartile deviation* (or, alternatively, the *semi-interquartile range*), i.e.:

$$Quartile\ deviation = \frac{Third\ quartile - First\ quartile}{2}$$

Example

Find the quartile deviation of the figures in Table 7D.
The first and third quartiles of Table 7D were calculated in 9:**14** as 433.3 and 474.2 kilometres respectively.

$$\therefore\ Quartile\ deviation = \frac{474.2 - 433.3}{2} = 20.45\ kilometres$$

3. Mean deviation. The *mean deviation* is simply the *average (mean) deviation of all the values from the distribution mean*. It is found, as its description would suggest, by adding up the deviations of all values from the distribution mean and dividing by the number of items. As a formula it can be written as:

$$Mean\ deviation = \frac{\Sigma(x - \bar{x})}{n}$$

(Note, however, that the sign of $x - \bar{x}$ must be ignored, i.e. all deviations are written as $+$.)

Example

Find the mean deviation of 5, 7, 8, 12 and 18, i.e. the figures used in 10:**5**.

$$\bar{x} = (5 + 7 + 8 + 12 + 18)/5 = 10$$

x	$x - \bar{x}$ (i.e. 10)
5	5
7	3
8	2
12	2
18	8
	20

$$\therefore\ Mean\ deviation = \frac{20}{5} = 4$$

The mean deviation is useful in giving some indication of the extent of the dispersion in terms of all the values in the distribution. It has not, however, any further statistical application and is not, therefore, a very important measure of dispersion.

4. Computing the mean deviation of a grouped frequency distribution. If the distribution is a grouped frequency distribution then the mean deviation is found as follows:

(*a*) Follow the normal procedure to find the mean – *see* 10:**5**.

(*b*) Add and complete the following two extra columns to the layout:

    (*i*) Difference: Class midpoint – mean (ignoring sign).

    (*ii*) Difference multiplied by class frequency.

(c) Add column (b) (ii) and divide by the total number of
frequencies of the distribution.

Example

Find the mean deviation of the distances in Table 7D.
Look back to 10:**5**.

Kilometres	f	Class mid-point (MP)	f×MP	Difference: MP − $\bar{x}$	f×(MP − $\bar{x}$)
400–under 420	12	410	4,920	44.5	534.0
420–under 440	27	430	11,610	24.5	661.5
440–under 460	34	450	15,300	4.5	153.0
460–under 480	24	470	11,280	15.5	372.0
480–under 500	15	490	7,350	35.5	532.5
500–under 520	8	510	4,080	55.5	444.0
Total	120		54,540		2697.0

∴ $\bar{x}$ = 54,540/120 = 454.5km   ∴ Mean deviation = 2697/120 = 22.475km

5. Standard deviation. The *standard deviation* (symbolised as *s*) is
found by adding the squares of the deviations of the individual values
from the mean of the distribution, dividing this sum by the number of
items in the distribution, and then finding the square root of the
quotient. Algebraically this can be shown as:

$$s = \sqrt{\left(\frac{\Sigma(x - \bar{x})^2}{n} \right)}$$

The standard deviation is by far the most important of the measures
of dispersion, but its importance is due to its mathematical properties
(especially in sampling theory – *see* part Six) rather than its descriptive
properties, i.e. its ability to make a distribution more easily
understood. However, note that the more the values of individual
items differ from the mean, the greater will be the square of these
differences and therefore the greater sum of the squares. And the
greater this sum, of course, the larger will *s* be. Hence the greater the
dispersion, the larger the standard deviation will be. (Note that if
there is no dispersion at all, i.e. all the values are the same, then the
standard deviation will work out at zero.)

But on the whole this is about as far as we can go in interpreting a
given standard deviation figure, though if the distribution is

symmetrical having a single frequency curve peak in the centre then it may well be that we can say some two-thirds of the distribution will fall in the range one standard deviation either side the mean.

6. Computation of the standard deviation. The standard deviation can, of course, be computed directly from the formula given in **5** above. The method is referred to as the *direct method*.

Example

To illustrate the direct method of finding s we will calculate the standard deviation of the figures previously used in **3**, i.e. 5, 7, 8, 12 and 18.

x	$(x - \bar{x})$	$(x - \bar{x})^2$
5	$5 - 10 = -5$	25
7	$7 - 10 = -3$	9
8	$8 - 10 = -2$	4
12	$12 - 10 = +2$	4
18	$18 - 10 = +8$	64
$\Sigma x = 50$		$\Sigma(x - \bar{x})^2 = 106$

$$\therefore \bar{x} = 50/5 = 10$$

$$s = \sqrt{\left(\frac{\Sigma(x - \bar{x})^2}{n} \right)}$$

$$= \sqrt{\left(\frac{106}{5} \right)} = \sqrt{21.2} = 4.6$$

If the mean is not a round number then the deviations will be in decimals. In such a case it is often easier to adopt an indirect method which uses the formula:

$$s = \sqrt{\Sigma x^2/n - \bar{x}^2}$$

Example

The above example will be reworked using the indirect method.

Now $\Sigma x^2 = 5^2 + 7^2 + 8^2 + 12^2 + 18^2 = 606$

So $s = \sqrt{606/5 - 10^2} = 4.6$

7. Computing the standard deviation from a grouped frequency distribution. As in the case of the mean, if the data is in the form of a grouped frequency distribution the 'direct method' cannot be used. To compute the standard deviation in this case the following procedure must be used:

(a) Follow the normal procedure to find the mean.

(b) Add and complete the following two extra columns to the layout:

    (i) the class midpoint squared;
    (ii) the class frequency multiplied by the class midpoint squared.

(c) Total column (b) (ii) to give Σx^2.

(d) Apply the formula:

$$s = \sqrt{\Sigma x^2/n - \bar{x}^2}$$

Example

Find the standard deviation of the distribution of distances in Table 7D. Look back to 10:5.

Kilometres	f	Class mid-point (MP)	$f \times MP$	MP^2	$x^2 (= f \times MP^2)$
400–under 420	12	410	4,920	168,100	2,017,200
420–under 440	27	430	11,610	184,900	4,992,300
440–under 460	34	450	15,300	202,500	6,885,000
460–under 480	24	470	11,280	220,900	5,301,600
480–under 500	15	490	7,350	240,100	3,601,500
500–under 520	8	510	4,080	260,100	2,080,800
Σ	120		54,540		24,878,400

$\therefore \ \bar{x} = 54{,}540/120 = 454.5\text{km}$

$$\therefore \ s = \sqrt{24{,}878{,}400/120 - 454.5^2}$$
$$= \sqrt{207{,}320 - 206{,}570.25}$$
$$= \sqrt{749.75} = \underline{\underline{27.38\text{km}}}$$

(Note that only one extra column need be added on the right-hand side of the layout if the user is prepared to square the class midpoint and multiply by the class frequency in a single operation.)

8. Symbols for population and sample means and standard deviations. At this juncture the reader is warned that it is conventional today to regard all sets of data as being samples, whether or not a population has been defined. In other words, although our 120 salesmen are all the items about which we wish to obtain information – and hence form our population (*see* 3:**1**) – they are regarded as a sample of a bigger population. This bigger population remains undefined – it could be all the salesman in the country, or the world, or all the drivers on company business. This approach somewhat blurs the distinction between a population and a sample but nevertheless is now the accepted practice.

The point about all this is that different symbols are used for the means and standard deviations of samples as against populations. Thus, $\bar{x}$ and s symbolise the mean and standard deviation of a *sample* while μ and σ symbolise the mean and standard deviation of a *population*. Since all raw data is deemed to form a sample, the former symbols are used in this part of the book, μ and σ being reserved until sampling theory is discussed later.

9. Variance. Variance is the name given to the square of the standard deviation, i.e. variance $= s^2$. It is important because variances can be added: for instance, if two independent distributions had variances of s_1^2 and s_2^2 respectively, the variance of the two distributions combined would be $s_1^2 + s_2^2$.

10. Units of the measures of dispersion. It should not be forgotten that the measures of dispersion are in the same units as the variable measured (*see* Chap. 9). For example, in Table 7D the units are kilometres. The range, quartile deviation and standard deviation are therefore all in kilometres. Had the problem involved a frequency distribution relating to ages of people (measured in the years) the measures would all have been in years.

11. Advantages and disadvantages of the different measures of dispersion. The following are the advantages and disadvantages of the measures of dispersion.

(*a*) *Range.*
  (*i*) Very simple to calculate.

(*ii*) Very simple to understand.

(*iii*) Used in practice as a measure of dispersion in connection with statistical quality control.

But . . .

(*iv*) Liable to mislead if unrepresentative extreme values occur.

(*v*) Fails totally to indicate the degree of clustering, e.g. 3, 5, 5, 5, 5, 7 has the same range as 3, 3, 3, 7, 7, 7, but in the former group values cluster much more closely together.

(*b*) *Quartile deviation*.

(*i*) Simple to understand.

But . . .

(*ii*) Fails to take into account all the values.

(*iii*) Gives no real indication of the degree of clustering.

(*c*) *Mean deviation*.

(*i*) Indicates the average dispersion of values from the mean.

(*ii*) Uses all the values in the distribution.

But . . .

(*iii*) Cannot be used in further statistical analysis.

(*d*) *Standard deviation*.

(*i*) Is of greatest importance in later statistical work (*see* Part Six).

(*ii*) Uses every value in the distribution.

But . . .

(*iii*) Difficult to comprehend as a descriptive measure.

(*iv*) Gives more than proportional weight to extreme values because it squares the deviations, e.g. a value twice as far from the mean as another is multiplied by a factor of four -2^2- relative to the latter value.

12. Relative dispersion: coefficient of variation. It sometimes happens that we need to compare the relative variability of two or more sets of figures. For example, are the figures in Table 7D relatively more variable than those given in **6**? The standard deviations are respectively 27 kilometres and 4.6 units. But these figures are clearly not comparable since, first, they are in different units and, second, they relate to sets of figures of quite different orders of size.

However, we could obtain some idea of the degree of the relative variability, or the relative dispersion, if we could relate the size of a variation to the average of the figures it was derived from, i.e. calculate the standard deviation as a percentage of the mean. This measure is called the *coefficient of variation*. It is expressed by the formula:

Coefficient of variation $= 100 \times s / \bar{x}$

To compare the variability of two sets of figures would therefore involve comparing their respective coefficients of variation.

Example

Compare the variability of the figures in Table 7D with those given in **6**.

	Table 7D	From **6**
s	27 km (*see* **7**)	4.6
$\bar{x}$	454.5 km (*see* **7**)	10

Coefficient of variation:

$(100 \times s / \bar{x})$	6	46

Conclusion: The figures used in **6** are very much more variable than those in Table 7D.

Skew

Finally, we consider the last descriptive measure of a given distribution, its *skew*.

13. Symmetry and skewness. If the histogram of a grouped frequency distribution is drawn, it usually displays quite low frequencies on the left, builds steadily up to a peak and then drops steadily down to low frequencies again on the right. If the peak is in the centre of the histogram and the slopes on either side are virtually equal to each other, the distribution is said to be *symmetrical* (*see* Fig. 11.1).

On the other hand, if the peak lies to one or other side of the centre of the histogram, the distribution is said to be *skewed* (*see* Figs 11.2 and 11.3). The further the peak lies from the centre of the histogram, the more the distribution is said to be skewed.

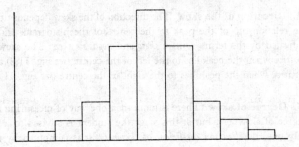

Figure 11.1 *Symmetrical distribution*

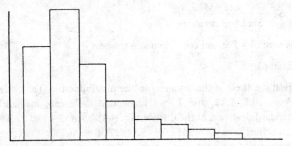

Figure 11.2 *Positive skew*

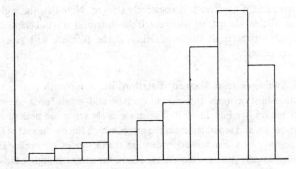

Figure 11.3 *Negative skew*

The skewness of a distribution can be measured as regards:

(a) the *direction* of the skew; and
(b) the *degree* of skew.

14. Direction of the skew. The direction of the skew depends upon the relationship of the peak to the centre of the histogram, and is indicated by the terms *positive skew* and *negative skew*. The skew is *positive* when the peak lies to the left of the centre (*see* Fig. 11.2) and *negative* when the peak lies to the right of the centre (*see* Fig. 11.3).

15. Degree of skew. There is more than one way of measuring the degree of skewness, but at this stage the student is advised to learn only one, the *Pearson coefficient of skewness*. It is computed by the formula:

$$Sk = \frac{3 \ (Mean - Median)}{Standard \ deviation}$$

Where Sk = Pearson coefficient of skewness.

Example

Find the skew of the distances in the distribution in Table 7D.
From 10:**5**, 9:**12**, and **7** we know that the mean, median and standard deviation of the distribution is 454.5, 452.4 and 27.38 km respectively. So $Sk = 3(454.5 - 452.4)/27.38 = +0.230$.

Incidentally, note that this formula automatically gives the *direction* of the skew, since if the answer is positive the distribution is positively skewed, and if negative it is negatively skewed. Note that the higher the coefficient the greater the skew. If the distribution is not skewed at all, but symmetrical, the application of the formula will give an answer of zero.

16. Averages in a skewed distribution. It is worth noting the relationship between the mean, median and mode of a skewed distribution (*see* Fig. 11.4(a)). The mode is always at the peak of the distribution and separated from the mean, which lies on the side of the longer tail, by a distance dependent on the degree of skewness. The median usually lies between the mode and the mean, though it can lie on the mode.

In the case of a symmetrical distribution the reader should note that the mean, median and mode all lie at the same point – at the centre of the distribution (*see* Fig. 11.4(*b*)).

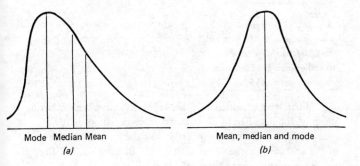

Figure 11.4 *Relationship between the mean, median, mode and skew.* (*a*) *Skewed distribution* (*b*) *Symmetrical distribution.*

Progress test 11

(*Answers in Appendix 4*)

1. Find the standard deviations and the mean deviations of the two distributors in Assignment 1 in Chapter 8. Which has the greatest relative variability?

2. (*a*) Compute the quartile deviation of the following travelling costs:

Daily travelling costs of office staff

Pounds per day	Number of staff
1–under 2	41
2–under 3	95
3–under 4	202
4–under 5	147
5 and over	15

(*b*) Also find the mean deviation and the standard deviation (no workings shown in suggested answers for this part of the question).

138 Frequency distributions

3. Find the coefficient of skewness for the following weekly branch sales:

Sales (£000)	Branches
Under 10	25
10–under 20	18
20–under 30	8
30–under 40	3
40 and over	1
	55

4. What is the coefficient of skewness of a distribution having a mean of 20, a median of 22 and a standard deviation of 10?

5. Without performing any further calculations, estimate the value of the arithmetic mean of the distribution in question 2 of Progress test 9 and justify your choice. (*CIMA*)

6. 'More than half our children are below average.' Comment.

Assignments

1. Find the range, quartile deviation, mean deviation, standard deviation and coefficient of skew for the data in the assignment in Chapter 7.

2. Find the ranges, quartile deviations, mean deviations, standard deviations and coefficients of skew for the distributions in Appendix 7. Comment on these measures.

3. (For calculators only.) First create a set of raw data by entering '8' into your calculator and then repeatedly multiplying by 1.02, so your first value is 8×1.02, your second 8×1.02×1.02, etc. (your first four values, then, will be 8.160, 8.323, 8.490 and 8.659, to three decimal places, and the last value 159.117). In this way form 151 values and from this raw data prepare an 8-class grouped frequency distribution. Then find, to three decimal places, the mean, median, mode, quartile deviation, mean deviation, standard deviation, coefficient of variation and Pearson's coefficient of skewness for this distribution.

(In the suggested answers these measures as computed from the raw data are given and by comparing your results with these you will be able to see if your answers broadly tally and the extent to which the loss of information on grouping has affected the accuracy of your measures. Since these differences will also be affected by how well judged your class limits were, it is suggested that you rework the assignment using other limits and seeing what effect this has on your answers.)

Part four
Correlation

12
Scattergraphs

Part three of this book was devoted to finding methods which described a *single* collection of figures. The objective in this part is the examination of *two* variables, i.e. two collections of figures, to see to what extent they are related. This kind of study is called *bi-variate analysis*. Three topics will be discussed: scattergraphs, regression lines and correlation. In order that the reader may compare these topics the two sets of figures shown in Table 12A will be used throughout for illustration.

Table 12A *Sales and advertising expenditure for Fred's mini-market 19–2 to 19–6*

Year	Advertising (£000)	Sales (£000)
	x	y
19–2	2	60
19–3	5	100
19–4	4	70
19–5	6	90
19–6	3	80

Relationship between two variables

If a car owner were to record daily the petrol he used and the kilometres he covered, he would find a very close relationship between the two sets of figures. As one increased, so would the other. On the other hand, if he compared his daily travelling with, say, the daily number of marriages in New York, he would find there was no relationship at all.

The relationship between kilometres driven and petrol used is an obvious one, but the relationship between other sets of figures is not usually so obvious. Businessmen have found by experience that there is a definite relationship between advertising and sales, but it is often difficult to say how close the relationship is. A study of the figures in Table 12A will reveal some connection, though an uncertain one, between advertising and sales in Fred's mini-market. Clearly some technique is called for to clarify this relationship.

1. Purpose of finding a relationship. Before considering such techniques the reader may rightly ask himself, why bother? The answer is that knowledge of the relationship enables us both to *estimate* and exercise *control*.

If we know there is a very close relationship between kilometres travelled and petrol used, it is possible to estimate how much petrol will be required for a given journey – or, conversely, how far it is possible to travel using a given quantity of petrol.

Alternatively, knowledge of the relationship can be used to control car performance, since if the distance obtained from a particular petrol consumption subsequently drops below what is expected it indicates that the engine is not functioning as it should. An overhaul will probably rectify matters and so enable the previous performance figures to be attained once more.

2. Estimates involving time lag. In business the ability to estimate one figure from another is particularly useful if there is a *time lag* between the two sets of figures.

For instance, there is a close relationship between the number of plans passed by a local authority in one year relating to houses to be built, and the number of baths bought the following year. It is possible for suppliers of baths, therefore, to estimate their next year's potential sales from this year's local government statistics (*see also* Appendix 4, answer 1 to Progress test 5, for a further example of time lag).

A similar search for relationships involving time lag goes on in the field of national and international economics, since a knowledge of such relationships enables economists to estimate the probable future economic position.

3. Independent and dependent variables. When dealing with two variables it is important to know which is the independent variable and which the dependent. The student is reminded (5:7) that:

The *independent* variable is the variable which is *not* affected by changes in the other variable.

The *dependent* variable is the variable which *is* affected by changes in the other.

In the case of Table 12A, changes in advertising for the year can be expected to affect sales. However, a change in the sales will not directly affect advertising expenditure. Thus, advertising is the independent variable and sales the dependent.

Sometimes it is not easy to decide which is which. In the case of the car owner above, is the independent variable petrol used or kilometres travelled? In practice, the answer to this sort of question often depends upon the way the data is collected. If predetermined quantities of petrol are put in the tank and the distance travelled is measured, then petrol used is the independent variable. If, on the other hand, predetermined distances are driven and the petrol consumption subsequently measured, then the independent variable is kilometres travelled.

Scattergraphs

The first technique we shall use to study the relationship between advertising and sales in Table 12A is the scattergraph. A *scattergraph* is a graph with a scale for each variable and upon which variable values are plotted in pairs.

4. Scattergraph construction. To construct a scattergraph:

(*a*) Prepare the graph so that the scale for the independent variable lies along the horizontal axis and the scale for the dependent variable lies on the vertical axis (*see* Fig. 12.1).

(*b*) Plot each pair of figures as a single point on the graph. Thus, in Table 12A, in 19−2, the £2,000 advertising and £60,000 sales form such a pair and therefore a point is plotted on the graph where the £2,000 line from the horizontal scale meets the £60,000 line from the vertical scale. This is really all there is to a scattergraph.

5. Purpose of a scattergraph. The basic purpose of a scattergraph is to enable us to see whether there is *any pattern among the points*. In the case of Table 12A data, it is clear there is some pattern, as the points tend to rise from left to right (*see* Fig. 12.1). The more distinct a pattern is, the more closely the two variables are related in some way.

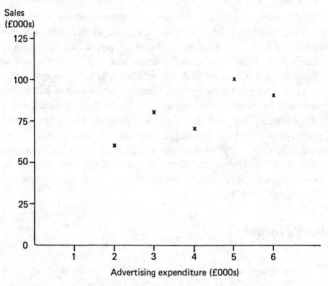

Figure 12.1 *Scattergraph of the Table 12A data*

6. The irrelevance of time. There is one point in connection with scattergraph construction that it is vital to appreciate, and that is, *time does not enter into the graph at all* (unless, that is, we are looking to see if there is a relationship between time and another variable – *see* Chapter 17). In the case of the data in Table 12A we are concerned with the relationship between advertising and sales, not *when* these amounts occurred. Thus the points on the scattergraph have no time significance whatever, the years in Table 12A merely linking specific advertising expenditure to specific sales achievements.

7. Line of best fit. If we now draw a straight line on the graph in such a way as to fit best the pattern of the points by lying as close to all the points as possible, then we have the *line of best fit* (*see* Fig. 12.2). This line is a valuable aid to understanding the relationship between the two variables, and the nearer the points lie to it the more valuable it is. When drawing the line of best fit, then, the object is to minimise the total divergence of the distances of the points from the line.

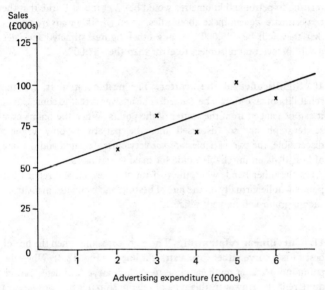

Figure 12.2 *Scattergraph of the Table 12A data with line of best fit added*

8. Estimates from a scattergraph. Once the line of best fit has been drawn, our scattergraph can be used for estimating simply by reading off from the line the sales value corresponding to any level of expenditure on advertising. Supposing £4,500 were to be spent on advertising; it can be seen from the line of best fit that this value is associated with sales of £85,000. Therefore £85,000 is the estimated sales to be obtained from an advertising expenditure of £4,500.

9. The reliability of estimates. The reader may not be too happy with this estimate. He may well have noticed from the scattergraph that sales nearly as large were once associated with advertising of only

£3,000, whereas on another occasion advertising expenditure of as much as £6,000 brought in only £90,000 worth of sales. How reliable, he may ask, is the estimate of £85,000?

The answer is 'Not very'. The relationship here between advertising and sales is just not close enough for estimates to be made with great accuracy (as, for example, the estimates discussed earlier relating to petrol and kilometres would be). Yet it is still true that the *best* estimate we can make about sales, given £4,500 of advertising, is that they will be £85,000, i.e. any other figure estimated for sales would be even more subject to error than the £85,000.

10. Significance of the scatter. The next question is, can the reliability of an estimate be gauged? The answer is that to some extent it can be gauged from the scatter of the points. When the points on a scattergraph are so dispersed that the pattern is only vaguely discernible, the two variables are *not* very closely related and the line of best fit is an unreliable guide for making estimates.

On the other hand, when the two variables are closely related the points will lie virtually on the line of best fit, and estimates made from such a graph will be very reliable.

11. Curvilinear relationship. There are occasions when the line of best fit is a curve rather than a straight line (*see* Fig. 12.3). When the points on a scattergraph lie close to such a curve, estimates can be quite reliable. However, this type of relationship (termed *curvilinear*) lies beyond the scope of this book and so will not be discussed further. All the relationships we shall consider will be straight-line, or *linear*, relationships. Before ever making a correlation analysis, therefore, the reader should check that this is so by drawing at least a rough scattergraph.

12. Limitations of scattergraphs. A thoughtful reader may by now have realised that scattergraphs have two serious limitations. These are:

(*a*) *Uncertainty as to the correct position of the line of best fit.* If the best estimates are to be made from the line of best fit, it must obviously first be drawn in the correct place. So far this has depended

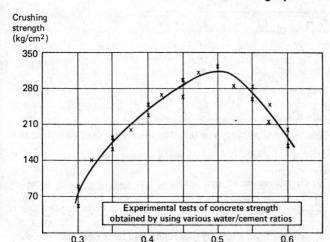

Figure 12.3 *Curvilinear relationship*

on the judgement of the person constructing the graph. It would be much better if a mathematical method of drawing the line could be devised instead of leaving it to the artistic whim of the individual.

(*b*) *Lack of a measure of the closeness of the relationship.* Since the reliability of an estimate depends heavily on the closeness of the relationship between variables, the lack of any measure of the closeness limits the value of scattergraphs considerably.

In the next two chapters we see how these limitations can be overcome.

Progress test 12

(*Answers in Appendix 4*)

1. The IQs of a group of six people were measured, and they then sat a certain examination. Their IQs and examination marks were as follows:

Person	IQ	Exam marks
A	110	70
B	100	60
C	140	80
D	120	60
E	80	10
F	90	20

Construct a scattergraph of this data, and draw the line of best fit.

(a) What marks do you estimate a candidate with an IQ of 130 would obtain?

(b) Estimate the IQ of a candidate who obtained a mark of 77.

2. Two dice were thrown and the sum of their pips doubled to give dice value. At the same time a card was drawn at random out of a normal pack from which all court cards had been removed. The value of this card was called the card value.

Throw	Dice value	Card value
1st	8	8
2nd	8	9
3rd	14	10
4th	22	5
5th	22	8
6th	16	3
7th	12	3
8th	6	2
9th	10	7
10th	10	5

Construct a scattergraph of this data. Is it possible to draw a line of best fit? What conclusions can you draw about the two variables?

Assignment

Create 50 pairs of numbers by starting with x at 777 and progressively adding 7, and y at 7 and progressively multiplying by 1.1 (rounding at each calculation to the nearest whole number). So the first three pairs will be: 777,7; 784,8; 791,9, and the last pair will be 1120 and 799. Draw a scattergraph of this data.

13
Regression lines

Note: It is not necessary to study this chapter to understand the rest of the book. Readers who are not required to study regression lines may turn at once to the next chapter.

At the end of the previous chapter it was pointed out that a line of best fit drawn by eye is dependent upon the subjective judgement of the person who draws it. Consequently the position of the line will differ slightly from person to person. To draw a line of best fit independent of individual judgement, its position has to be found mathematically. Such a line is called a *regression line*.

Computing regression lines

1. Equation of the line. The general equation for any straight line on a graph is

$$y = a + bx$$

where a and b are constants. Positioning a regression line means, therefore, finding the appropriate values of a and b.

Note: It is assumed that the reader fully understands this equation. If he does not then he should refer to Appendix 1 where the basic mathematical theory is outlined.

2. The method of least squares. When drawing a line of best fit, we said that an attempt is made to minimise the total divergence of the points from the line (*see* 12:**7**). In computing the line mathematically, the same idea is pursued, only it has been found that the best line is one that minimises the total of the *squared* deviations. This computation is logically known as the *method of least squares*.

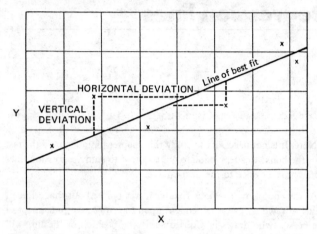

Figure 13.1 *Measuring horizontal and vertical deviations from the line of best fit*

3. Measuring the deviations. When it comes to measuring the deviations of the points from the line, it is important to understand that statisticians do *not* measure the shortest distance between a point and the line of best fit. What they measure is either:

(*a*) the *vertical* distance between point and line; or
(*b*) the *horizontal* distance (*see* Fig. 13.1).

These two different ways of measuring the deviations will produce *two different regression lines*, one minimising the total of the squared deviations measured vertically, the other minimising the total of the squared deviations measured horizontally.

This can be rather confusing, so for the moment only the calculations needed to determine the regression lines will be considered. The distinction between them will then be dealt with in **7** *et seq.*

First, let us take the regression line in which deviations are measured vertically. This is by far the commoner of the two.

4. The regression line of y on x. The regression line found by measuring the deviations vertically is called the *y on x regression line* and the *a* and *b* for this line are found from the following formulae:

$$b = \frac{n \times \Sigma xy - \Sigma x \times \Sigma y}{n \times \Sigma x^2 - (\Sigma x)^2}$$

$$a = \frac{\Sigma y - b \times \Sigma x}{n}$$

where *n* is the number of pairs of figures (points) and *x* and *y* are the variable values relating to the horizontal and vertical axes respectively.

Note: (*a*) Since the formula for *a* uses *b*, *b* has to be found first.
(*b*) In examinations you are sometimes given the 'Sum of the squares'. This is Σx^2 or Σy^2 (or whatever variable is involved).

To illustrate the use of the formulae the data of Table 12A will be analysed (it will be noticed that the advertising column is already headed *x* and the sales column *y*).

Example

Examination of the two formulae shows that the only figures required are Σy, Σx, Σxy, Σx^2 and *n*, *n* in this case being 5 as there are five pairs of figures. The other four figures can be quickly obtained by laying out the data in tabular form:

From Table 12A		Computed	
x	y	xy	x^2
2	60	120	4
5	100	500	25
4	70	280	16
6	90	540	36
3	80	240	9
$\Sigma x = 20$	$\Sigma y = 400$	$\Sigma xy = 1,680$	$\Sigma x^2 = 90$

It only remains now to insert these values into the two formulae and solve for a and b:

$$b = \frac{n \times \Sigma xy - \Sigma x \times \Sigma y}{n \times \Sigma x^2 - (\Sigma x)^2} = \frac{5 \times 1,680 - 20 \times 400}{5 \times 90 - 20^2} = \frac{8,400 - 8,000}{450 - 400} = \underline{\underline{8}}$$

$$a = \frac{\Sigma y - b \times \Sigma x}{n} = \frac{400 - 8 \times 20}{5} = \underline{\underline{48}}$$

The regression line, therefore, is $y = a + bx = 48 + 8x$.

This is the *regression line of y on x* and is the line of best fit when the deviations are measured vertically.

5. The regression line of x on y. If it is desired to compute the second regression line, where the deviations are measured horizontally, it is only necessary to alter the two formulae so that the xs and the ys are interchanged. The formulae, then, become:

$$b = \frac{n \times \Sigma yx - \Sigma y \times \Sigma x}{n \times \Sigma y^2 - (\Sigma y)^2}$$

$$a = \frac{\Sigma x - b \times \Sigma y}{n}$$

Note that in the b formula, since yx is the same as xy, the numerator is exactly the same as before – and, indeed, the only new figure that must be computed is Σy^2.

Example

Finding this second regression line in the case of our data in Table 12A, then, requires us only to compute Σy^2 before applying the formulae. And $\Sigma y^2 = 60^2 + 100^2 + 70^2 + 90^2 + 80^2 = 33,000$, so we have:

$$b = \frac{n \times \Sigma yx - \Sigma y \times \Sigma x}{n \times \Sigma y^2 - (\Sigma y)^2} = \frac{5 \times 1,680 - 400 \times 20}{5 \times 33,000 - 400^2} = \frac{400}{165,000 - 160,000} = \underline{\underline{0.08}}$$

$$a = \frac{\Sigma x - b \times \Sigma y}{n} = \frac{20 - 0.08 \times 400}{5} = \frac{20 - 32}{5} = \underline{\underline{-2.4}}$$

Therefore the equation of the line is $x = -2.4 + 0.08y$.

(Note that this equation is *not* merely the previous one turned round.)

And this line is known as the *regression line of x on y*, and is the line of best fit when the deviations are measured horizontally.

6. Graphing regression lines. It is quite easy to graph the regression lines once they have been computed. All one has to do is:

(*a*) choose any two values (preferably well apart) for the unknown variable on the right-hand side of the equation;

(*b*) compute the other variable values;

(*c*) plot the two pairs of values; and

(*d*) draw a straight line through the plotted points.

Example

Graph of regression lines for Table 12A as computed above.

(1) *Regression line of y on x*

(*a*) Let $x = £6,000$
$$\therefore y = 48 + 8 \times 6 \text{ (remember, } x \text{ was in '000s)}$$
$$= £96,000 \text{ (} y \text{ too was in '000s)}$$

(*b*) Let $x = £0$
$$\therefore y = 48 + 0$$
$$= £48,000$$

These points, and the regression line through them, are shown in Fig. 13.2.

(2) *Regression line of x on y*

(*a*) Let $y = £100,000$
$$\therefore x = -2.4 + 0.08 \times 100$$
$$= £5,600$$

(*b*) Let $y = £50,000$
$$\therefore x = -2.4 + 0.08 \times 50$$
$$= £1,600$$

Again, these points, and the regression line through them, are shown in Fig. 13.2.

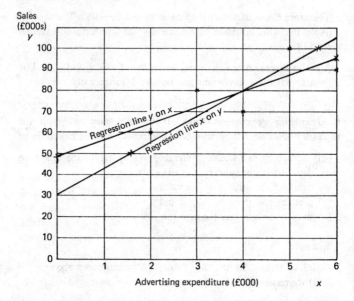

Figure 13.2 *Regression line of the Table 12A data.* A point of interest is that the *y* on *x* and the *x* on *y* regression lines always intersect at the means of the two series – here at £4,000 and £80,000.

The use of regression lines

7. Use of the regression line of *y* on *x*. Using the *y* on *x* regression line – the one first calculated in **4** – is quite simple. To revert to our illustrative figures (Table 12A) it is only necessary to replace *x* in the equation by a value for advertising to obtain an estimate of the sales, *y*. If we again take the advertising figure of £4,500 previously selected in **12:8**, the estimated sales will be:

$$y = 48 + 8 \times 4\frac{1}{2}$$
$$= 48 + 36 = 84, \text{ i.e. sales of £84,000.}$$

8. Use of the regression line of *x* on *y*. In a similar way the regression line of *x* on *y* is used to estimate the value of *x* that follows from any given value of *y*. In the example so far discussed, it means estimating the advertising that would be associated with any given sales value. If, for example, sales had been £84,000, the advertising estimated to have been incurred would be:

$$x = -2.4 + 0.08 \times 84$$
$$= 4.32, \text{ i.e. } £4,320$$

It will be seen that this is a different figure from the £4,500 advertising that gave an estimated £84,000 sales. This paradox is one of the reasons why regression lines seem so confusing to students.

9. The regression line paradox. In the previous paragraph an advertising expenditure of £4,320 was estimated to have been incurred, given a sales value of £84,000. Yet in the paragraph before that, £84,000 was the estimated sales value of £4,500 of advertising. Why isn't a sales value of £84,000 associated with a single advertising figure? The reason is not easy to grasp, but some insight into the underlying reason may be gathered from the following explanation (*see* Table 12A).

If the *highest* advertising value were incurred, would it be wise to predict a sales value equal to the highest recorded? Obviously not, since the highest advertising (£6,000) was associated with a sales value of only £90,000, i.e. £10,000 below the maximum. A better offhand estimate would probably be around the amount actually attained – £90,000. Now if sales were £90,000, and we were unaware of the actual level of advertising expenditure, would it be wise to estimate that the highest advertising had been incurred? Since, on one occasion, sales of £100,000 followed an advertising expenditure of only £5,000, this too would be foolish. An estimate of advertising below the maximum would be more sensible.

All this means is that *given* the maximum advertising, an *estimate* of below-maximum sales is made; but *given* this specific sales figure, an *estimate* of below-maximum advertising must be made. The given figure and the estimate *cannot be interchanged*. In other words, if an advertising figure of £i leads to an estimate of £j sales, you cannot, on learning that in a certain year sales of £j were made, estimate advertising to have been £i.

Regression lines are like one-way streets: you can only move in the authorised direction. If you want to reverse direction you must move across to the other regression line.

10. Choice of regression line. We have seen that there are two regression lines and that it is important to choose the right one. The next problem is *how* to choose the right one. Luckily the answer is quite simple: always use the line that has *the variable to be estimated on the left-hand side of the equation of the line.* If the y variable is to be estimated, use the $y = a + bx$ (y on x) line; if x is to be estimated, use the $x = a + by$ (x on y) line.

Indeed, remembering which line is which can perhaps be made easier still by noting the following left-hand/right-hand patterns of the regression line formulae:

Left-hand   *Right-hand*

Estimate y given x:

Use $y = a + bx$ equation,

which gives the y on x regression line.

Left-hand   *Right-hand*

Estimate x given y:

Use $x = a + by$ equation,

which gives the x on y regression line.

11. Reliability of estimates. Although we have just been through a quite lengthy procedure for obtaining estimated values, it is important to realise that such estimates are no more reliable than those derived from a well constructed scattergraph. The procedure simply ensures that the scattergraph *is* well constructed. The reliability of estimates depends far more on the closeness of the relationship between the variables than on the most elaborate mathematical calculations. The immediate reason for computing a regression line (other than to obviate the possibility of a badly judged line) is to obtain a line of best fit free of subjective judgement.

A more important reason, though it does not concern us here, is that in advanced statistics the reliability of estimates made from such a line can be measured mathematically.

12. Regression coefficient. The term regression coefficient occasionally arises in statistical discussion. This is simply the value of b in the equations just discussed. For example, in the case of the regression of sales on advertising, the regression coefficient was 8 and in the case of the regression of advertising on sales it was 0.08.

Progress test 13

(*Answers in Appendix 4*)

1. Find the regression lines relating to the data in question 1 of Progress test 12 and estimate (*a*) the marks that would be obtained by a candidate with an IQ of 130, (*b*) the IQ of a candidate who obtained 77 marks.

Superimpose these lines on the scattergraph constructed for question 1 in Progress test 12.

2. Using the data in question 2 in Progress test 12.

(*a*) compute the regression line of card value on dice value and use this line to determine the best estimate of the card value that would be associated with a dice value of (*i*) 4, (*ii*) 24;

(*b*) compute the regression line of dice value on card value and use the line to determine the best estimate of the dice value that would be associated with a card value of (*i*) 1, (*ii*) 10;

(*c*) using this example as an illustration explain why it is necessary to compute two regression lines, one for estimating a y value from an x value and the other an x value from a y value?

Superimpose these regression lines on the scattergraph constructed for question 2 in Progress test 12.

3. Steel bars of rectangular cross-section are forged into bars of smaller cross-section. The time to forge a bar is found to be approximately proportional to the difference between the cross-sectional area (CSA) at the start and finish of forging, plus a fixed time which is independent of the cross-sectional areas. That is,

$t = a + b$ (CSA at start − CSA at finish)

The following forging times were observed for a sample of bars:

Starting size (inches)	Finishing size (inches)	Forging time (minutes)
9×4	4×1	16
8×6	4×2	24
4×4	1×1	9
8×2	2×2	12
8×8	4×4	29
7×7	5×2	21
6×7	5×2	14

Required:

(a) Find, using least squares regression, the best fit relationship of the above type.

(b) Use your regression equation found above to

    (i) predict how much longer it would take to forge a bar 2 inches × 3 inches from a bar 6×6 inches square than from a bar 5×5 inches square;

    (ii) estimate the CSA of the finished size of a bar if the starting CSA is 36 sq. inches and the forging time is 18 minutes.

(*ACCA*)

Assignment

Find the *x on y* and the *y on x* regression lines of the data in the assignment in Chapter 12.

14
Correlation

We consider next the second limitation of scattergraphs mentioned at the end of Chapter 12. There it was pointed out that the reliability of an estimate depended on the closeness of the relationship between the two variables and that a measure of this closeness was essential for assessing reliability. There are different measures of this closeness, but the most generally used is one called the *Pearson product-moment coefficient of correlation*, commonly symbolised as *r*.

Computation of *r*

First we look at how *r* is computed though before we start the reader is reminded of the qualification made in 12:**11**, namely that in this book we are only concerned with linear correlation and all the theory assumes such linearity.

1. Formula for *r*. The formula for calculating *r* can be expressed in a number of ways. Perhaps the most practical is:

$$r = \frac{n \times \Sigma xy - \Sigma x \times \Sigma y}{\sqrt{(n \times \Sigma x^2 - (\Sigma x)^2)(n \times \Sigma y^2 - (\Sigma y)^2)}}$$

Example

Find r of the data in Table 12A.

First we find the various Σ values (and at this point the reader should note the similarity of the computational layout and the *b* formula in the previous chapter to the computational layout and the *r* formula in this).

From Table 12A			Computations	
x	y	x^2	y^2	xy
2	60	4	3600	120
5	100	25	10000	500
4	70	16	4900	280
6	90	36	8100	540
3	80	9	6400	240
$\Sigma 20$	400	90	33000	1680

Then we apply the formula:

$$r = \frac{n \times \Sigma xy - \Sigma x \times \Sigma y}{\sqrt{(n \times \Sigma x^2 - (\Sigma x)^2)(n \times \Sigma y^2 - (\Sigma y)^2)}}$$

$$= \frac{5 \times 1680 - 20 \times 400}{\sqrt{(5 \times 90 - 20^2)(5 \times 33000 - 400^2)}} = \frac{8400 - 8000}{\sqrt{50 \times 5000}} = +0.80$$

Note that r is not expressed in any units.

Interpretation of r

It has been repeatedly emphasised that the reliability of estimates depends upon the closeness of the relationship between two sets of figures. Now that we have a measurement of this closeness, the student is probably keen to learn how to interpret this figure, particularly in assessing the reliability of estimates.

Unfortunately, such interpretation depends very much on experience. The full significance of r will be grasped only after working on a number of correlation problems and seeing the kinds of data which give rise to various values of r. Until this experience has been gained the student would be wise to interpret r very cautiously and to restrict such interpretation to the most general terms. However, to give him a little insight into the significance of r, its interpretation will now be discussed in just such general terms.

2. Phrases to describe correlation. Before we examine the numerical significance of r it is necessary to define certain phrases that are commonly used to describe correlation. (Even better than verbal description are actual examples shown on a scattergraph, and the reader should match the definitions below against the illustrations in Fig. 14.1.)

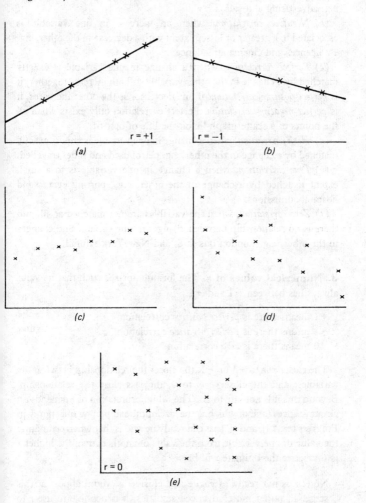

Figure 14:1 *Types of correlation.* (a) Perfect positive correlation. (b) Perfect negative correlation. (c) High positive correlation. (d) Low negative correlation. (e) Zero correlation, variables uncorrelated.

(*a*) *Positive correlation*, when an increase in one variable is associated to a greater or lesser extent with an increase in the other, e.g. advertising and sales.

(*b*) *Negative correlation*, when an increase in one variable is associated to a greater or lesser extent with a decrease in the other, e.g. TV licences and cinema attendance.

(*c*) *Perfect correlation*, when a change in one variable is exactly matched by a change in the other variable. If both increase together, it is *perfect positive correlation*: if one decreases as the other increases, it is *perfect negative correlation*. Perfect correlation only exists when all the points of a scattergraph lie on the line of best fit.

(*d*) *High correlation*, when a change in one variable is almost exactly matched by a change in the other, e.g. petrol used and miles travelled.

(*e*) *Low correlation*, when a change in one variable is to a small extent matched by a change in the other, e.g. sporting events and hospital admissions.

(*f*) *Zero correlation*, when the variables are not matched at all, and there is no relationship between changes in one variable and changes in the other, e.g. London bus fares and New York rainfall.

3. Numerical values of r. The formula for r is such that its value always lies between -1 and $+1$.

$+1$ means there is *perfect positive* correlation.
-1 means there is *perfect negative* correlation.
0 means there is *zero* correlation.

The closer r is to $+1$ or -1, the closer the relationship between the variables; and the closer r is to 0, the less close the relationship. Beyond this it is not safe to go. The full interpretation of r depends on circumstances (one of which is the size of the sample: *see* question 5 in Progress test 14), and all that can really be said is that when estimating the value of one variable from the value of another, then the higher r is the better the estimate will be.

Note: r is not really as vague and elusive as would appear at this stage. In rather more advanced statistics, it is possible to use r to indicate quantitatively the reliability of an estimate.

One final point should be made. The closeness of the relationship is not proportional to r: an r of (say) 0.8 does *not* indicate a relationship twice as close as one of 0.4 (it is, in fact, very much closer).

4. Spurious correlation. When interpreting r it is vitally important to realise that there *may be no direct connection at all between highly correlated variables*. When this is so the correlation is termed *spurious* or *nonsense* correlation. It can arise in two ways.

(*a*) There may be an *indirect connection*. For example, motorway driving and holidays on the Continent are probably quite highly correlated, since both tend to increase with a rising standard of living. To draw the conclusion that motorway driving *causes* Continental holidays or vice versa would be quite wrong, and a decision (say) by the Chancellor of the Exchequer to close Britain's motorways as part of a campaign to reduce British expenditure abroad would be ridiculous.

(*b*) There may be a *series of coincidences*. Normally, spurious correlation cannot arise in the course of properly conducted statistical work. Laymen, however, sometimes fall into the following trap. They examine the series relating to a variable they wish to predict – say the weather in June – and then hunt about amidst all sorts of data looking for a series that correlates highly with this variable. Needless to say, it is quite likely that, by chance, one set of data out of the hundreds at hand will fit, even if it is something as unlikely as the number of divorces in January. Then they draw everyone's attention to the high correlation and proceed to predict next June's weather on the basis of it. Such predictions are, of course, quite worthless.

5. Correlation and regression lines. Those who read Chapter 13 will probably be interested to know that the higher the correlation, the smaller the angle between the two regression lines. In the case of perfect correlation (positive or negative) the two lines will coincide and there will be only one line.

Conversely, the lower the correlation, the greater the angle between the two lines, until, when r is zero, the lines lie at right angles to each other – the *y on x* line being horizontal and the *x on y* line vertical.

This, of course, is quite logical. Remember that the two regression lines intersect at the means of the two series. So if the lines are horizontal and vertical then the horizontal line will fall along the mean of the y series and the vertical line will fall along the mean of the x series. Now, if two variables are wholly unconnected then no matter what the value of one is, the best estimate of the other is the mean of that other series. Using the horizontal and vertical regression lines to estimate in this situation results in just those means being estimated.

Rank correlation

In some sets of data the actual values are not given, only the order in which the items are ranked.

Example

The examination places of a class of seven boys taking examinations in physics and French are as follows:

Boy	Physics	French
Allen	3	1
Birch	2	4
Clarke	1	2
Davis	4	3
Evans	6	5
Ford	5	7
Gregory	7	6

In this situation we cannot calculate r (the rankings do not tell us by just *how much* a person in rank 1 leads the person in rank 2 and the person in rank 2 leads the person in rank 3). However, there is a less precise but still useful measure of correlation we can compute. This is called *rank correlation* and it is this we look at now.

6. Computation of rank correlation. Rank correlation is the measure of correlation we obtain by using the r formula in **1** and treating the rankings as variable values (e.g. rank 5 in the x series means x takes the value '5'). However, because the difference between adjacent ranks is always 1, an alternative and simpler formula (which gives what is called *Spearman's coefficient of rank correlation*) is:

$$\text{Spearman's coefficient of rank correlation} = 1 - \frac{6\Sigma d^2}{n(n^2 - 1)}$$

where d = the *difference* between the rankings of the same item in each series (since these differences are squared in the formula there is no need to bother about whether they are positive or negative).

The computation of Spearman's coefficient of rank correlation in the above example is, therefore:

Physics Rank	French Rank	d	d^2
3	1	2	4
2	4	2	4
1	2	1	1
4	3	1	1
6	5	1	1
5	7	2	4
7	6	1	1
			16

Applying formula

$$\text{Spearman's coefficient of rank correlation} = 1 - \frac{6\Sigma d^2}{n(n^2 - 1)}$$

$$= 1 - \frac{6 \times 16}{7(7^2 - 1)} = 0.714$$

7. Tied ranking. A minor problem arises when two items in one or other of the series tie in rankings, e.g. Clarke and Davis may be third equal in French. Note that in such a situation the next rank jumps two, not one, i.e. Evans in the example just given would still rank fifth, not fourth. When you have tied rankings it means that two (or even more) items share the next rankings between them. To reflect this, the rankings involved are added and the sum divided by the number of items sharing, e.g. if Clarke and Davis are third equal then they share ranks three and four and each is given a ranking of $(3+4)/2 = 3\frac{1}{2}$. And if W, X, Y and Z share rankings of 7, 8, 9 and 10 they would each be given a ranking of $(7+8+9+10)/4 = 8\frac{1}{2}$.

Once the tied rankings have been set against the sharing items, the computation of Spearman's coefficient of rank correlation is made in the usual way.

8. Interpretation of Spearman's coefficient of rank correlation. Spearman's coefficient of rank correlation is interpreted in exactly the same way as r. However, it should be appreciated that since one has less information when only the rankings are given then the coefficient is not as informative as it would be if it were calculated from the *measured* variables for the two series (e.g. from the full sets of marks for both examinations).

Progress test 14

(Answers in Appendix 4)

1. Compute the coefficient of correlation for the data in question 1 in Progress test 12.
2. Compute the coefficient of correlation for the data in question 2 in Progress test 12 and state what type of correlation exists between the two sets of figures.
3. Lay out the data in question 1 in Progress test 12 in the form of rankings and compute the rank correlation. Why does the correlation figure obtained not equal the figure obtained in the answer to question 1?
4. The finance division of a large company is investigating its procedures for the selection of new accountancy trainees. Potential applicants are given prior to appointment both a written test and a formal interview. The performances of eight successful applicants were rated after their first full year with the company. The independent rankings of written test, interview assessment and job performance for the eight trainees are given below:

(1 = best, 2 = second best, etc.)

Trainee	A	B	C	D	E	F	G	H
Written test	6	2	7	4	1	5	3	8
Interview	1	4	2	3	6	5	8	7
Job performance	1	2	3	4	5	6	7	8

$$\left[\text{Spearman's } r_s = 1 - \frac{6\Sigma d^2}{n(n^2 - 1)} \right]$$

You are required to:

(a) calculate a rank correlation coefficient between:
  (i) job performance and written test;
  (ii) job performance and interview assessment;
(b) interpret your results.

(*CIMA*)

5. Determine the coefficient of correlation for the data in the following:

Year	x	y
19–5	5	17
19–6	75	19

Explain why your answer comes to the figure it does. What conclusion can you draw from this exercise regarding the interpretation of *r* in small samples?

6. In the course of a survey relating to examination success, you have discovered a high *negative* correlation between students' hours of study and their examination marks. This is so at variance with common sense that it has been suggested an error has been made. Do you agree?

Assignments

1. Find the coefficient of correlation of the data in the assignment in Chapter 12.

2. Rank the figures in the two series in the assignment in Chapter 12 and find the ranked correlation. What does this tell you about Spearman's coefficient of rank correlation?

Part five
Other descriptive statistical topics

15
Index numbers

In Part three the function of statistics in reducing a mass of data, with the aid of measures, to a form easier to grasp was emphasised. We now return to this aspect of statistics, this time seeing how single figures called index numbers can summarise masses of data.

The theory of index numbers

In the first part of Chapter 10 it was pointed out that using a single figure to represent a host of others results in many different opinions as to which figure serves best. With index numbers this feature is again present so that there emerges a confusing variety of types of index numbers and methods of calculation. Moreover, the very act of compromise inherent in selecting one figure to represent many others inevitably results in doubt as to the worth of the index number even when it has been obtained. In this chapter only the minimum number of types will be examined and their worth left unexamined.

1. The concept of an index number. If we wish to compare several series of figure it is more than likely that their complexity will render direct comparison meaningless. If, for instance, we had information on every form of production during this year and last year, e.g. tonnes of steel produced, litres of paint blended, TV sets manufactured, cars assembled and so on, the sheer mass of data would make it impossible to 'see' in which year production was higher. Instead of such an embarrassing excess of figures, what we need is a *single* figure which in itself shows how much one year differs from another. A convenient way of doing it is to take a fairly typical year's figures as a base, and express the figures for other years as a percentage of this. Hence, if the figure for 19−5 were 100 and that for 19−6 105, we should know that production (or whatever) was 5 per cent greater.

Such a single figure summarising a comparison between two sets of figures is called an *index number*.

2. Base 100 – the percentage relative index. In many ordinary day-to-day comparisons 100 is used as a base, percentages being the most obvious example. In consequence, people have become used to such comparisons and statisticians take advantage of this fact by basing index numbers on 100.

For example, if the production of TV sets was 38,261 last year and 43,911 this year we could call last year's production 100 and this year's (by simple proportion) 115. In this way the comparison between the two years' production is made much clearer. Note that an index number computed in this way is called a *percentage relative index*.

A few indices use other bases, such as 10 or 1,000, but they are exceptional and there are usually good reasons for such a departure from normal practice.

3. Base year. When comparing a series of annual figures (by far the commonest application of index numbers) it is necessary first to select one of the years as a base and designate the figure relating to that year as 100. Then all the other figures are expressed in terms of this selected year. A *base year* can, therefore, be defined as *the year against which all other years are compared*.

Sometimes a base year alters as the series progresses (e.g. as in the case of a chain index, *see* **17**) – but normally once chosen, a base year remains unchanged for the whole of the series.

4. Index number symbols. All the different methods of calculating index numbers can be expressed concisely and unambiguously as formulae. The symbols used in these formulae are the following:

p price of individual items;
p_0 price of individual items in base year;
p_1, p_2, p_3, etc.   prices of individual items in subsequent years*;

q quantity of individual items;
q_0 quantity of individual items in base year;
q_1, q_2, etc. quantities of individual items in subsequent years*;
w weight.

Note that Σpq means that the price and quantity of each item in turn are first multiplied together and the products then added.

*If the current year only is being compared with base year, then the suffix 1 indicates the current year.

5. One-item index numbers. Where only one item is involved in comparisons between different periods, the calculation of index numbers is very simple. One year is chosen as base, and the values for other years are stated in proportion to the value of the base year, i.e.:

$$\text{Quantity index} = \frac{q_1}{q_0} \times 100$$

or

$$\text{Price index} = \frac{p_1}{p_0} \times 100$$

Example

With 19–5 as base year, compute quantity and price indices for the years 19–3 to 19–9.

Year	Price (£)	TV sets sold	Price index		Quantity index	
19–3	450	12,912	$\frac{450}{500} \times 100 =$	90	$\frac{12,912}{21,200} =$	61
19–4	480	18,671	$\frac{480}{500} \times 100 =$	96	$\frac{18,671}{21,200} =$	88
19–5	**500**	**21,200**	etc.	**100**	etc.	**100**
19–6	530	28,633		106		135
19–7	530	35,028		106		165
19–8	550	40,650		110		192
19–9	600	44,531		120		210

Note that a one-item price index is also called a *price relative, see* **15**.

Weighted aggregative indices

Unfortunately, index numbers are often wanted in circumstances where there is more than just one item. To take a common instance, we often need an index number that compares the cost of living in one year with that in another. Clearly, more than one item is involved in the cost of living!

For illustrative purposes, let us take a fictitious country in which only three items enter a cost of living index – bread, cheese and ale – and that prices in the two years to be considered were as indicated in Table 15A.

Table 15A *Data for three-item cost of living index*

Item	19–0	19–5
Bread	45p loaf	50p loaf
Cheese	500p kg	1,000p kg
Ale	900p keg	750p keg

6. Difficulties involved in multi-item indices. Our aim now is to determine a single figure which will compare the cost of living in 19–5 with that of 19–0. Examination of the figures above reveals at least three difficulties.

(*a*) Two prices have gone up and one down. As there can only be a single index number *it must be a compromise* between these two opposing price movements.

(*b*) The prices are given *for different units*. It is not feasible, therefore, to add together all the prices for a single year.

(*c*) There is no indication as to *how important* each item is in the cost of living. Possibly bread should be considered more important than ale.

Difficulty (*a*) is a feature of index numbers that must always be borne in mind. Index numbers *are* compromises.

7. Weighting. Difficulties **6**(*b*) and (*c*), on the other hand, can be overcome by *weighting*, i.e. multiplying the price by a number (the *weight*) that will adjust the item's value in proportion to its importance. For example, if cheese is given a weight of 10, its original price of 500p becomes a weighted price of 5,000p.

When such a weight is selected, both the importance of the item and the unit in which the price is expressed are taken into consideration. Consequently, weighted figures are directly comparable.

Assume that weights of 100, 10 and 1 are given respectively to bread, cheese and ale. The weighted figures will therefore be as indicated in Table 15B.

Table 15B *Weighted cost of living index*

Item	Weight	19-0 Price	19-0 Price ×weight	19-5 Price	19-5 Price ×weight
Bread	100	45p	4,500	50p	5,000
Cheese	10	500p	5,000	1,000p	10,000
Ale	1	900p	900	750p	750
Total			10,400		15,750

From the completed table it is possible to compute a single index number simply by calling the total of the 19-0 weighted price column '100', and finding the total of the 19-5 column as a proportion, i.e.:

$$\text{Index number for 19-5 (19-0=100)}=\frac{15,750}{10,400}\times 100 = \underline{\underline{151}}$$

8. Summary of procedure – weighted aggregative price index. The procedure we adopted can be summarised as follows:

(a) list the items and prices;
(b) select weights;
(c) multiply the prices by selected weights (*weight*);
(d) add the products (*aggregate*);
(e) compare the total for the base year with the total for the other year by using a percentage (*index*).

From this summary it will be clear why an index computed by this method is called a *weighted aggregative price index*.

9. Formula for a weighted aggregative price index. The formula for calculating a weighted aggregative price index is:

$$Index=\frac{\Sigma(p_1\times w)}{\Sigma(p_0\times w)}\times 100$$

Laspeyre and Paasche indices

Probably in practice the commonest indices met with are the Laspeyre and the Paasche.

10. Quantity weighted indices. In the previous section the actual selection of weights for a weighted aggregative index was not discussed. In practice this can be a difficult problem. One solution is to use the *actual quantities consumed* as weights. Obviously , the more bread that is consumed the more important bread is as a cost of living item. Similarly, the low consumption of, say, caviare would reflect the insignificance of such an item. An index number so computed is known as a *quantity weighted index*. Such an index differs from the one discussed above only in the use of actual quantities for weights.

11. Base year or current year quantities? Using actual quantities is all very well, but the question immediately arises *which* quantities – those consumed in the base year, or those consumed in the year for which the index is required?

The answer is that either can be used, although, of course, different index numbers are obtained as a result. It so happens that each method is named after its original inventor. The one which uses base year quantities is called a *Laspeyre* index and the one that uses the current year quantities a *Paasche* index.

12. Definitions and formula.

(*a*) *Laspeyre price index.* This is a base year quantity weighted index. The formula is:

$$Index = \frac{\Sigma p_1 q_0}{\Sigma p_0 q_0} \times 100$$

A Laspeyre price index indicates how much the cost of buying base year quantities at current year prices is, compared with base year costs.

(*b*) *Paasche price index.* This is a current year quantity weighted index. The formula is:

$$Index = \frac{\Sigma p_1 q_1}{\Sigma p_0 q_1} \times 100$$

A Paasche price index indicates how much current year costs are compared to the cost of buying current year quantities at base year prices.

Example

Compute (1) *Laspeyre and* (2) *Paasche price indices of the following data* (*19–0 = base year*):

Item	19–0		19–5	
	Price (p_0)	Quantity (q_0)	Price (p_1)	Quantity (q_1)
Bread	45p loaf	80,000 loaves	50p loaf	100,000 loaves
Cheese	500p kg	10,000 kg	1,000p kg	15,000 kg
Ale	900p keg	1,000 kegs	750p keg	3,000 kegs

(1) *Laspeyre price index:*

Item	p_0	p_1	q_0	$p_0 \times q_0$ (£)	$p_1 \times q_0$ (£)
Bread	45p	50p	80,000	36,000	40,000
Cheese	500p	1,000p	10,000	50,000	100,000
Ale	900p	750p	1,000	9,000	7,500
				95,000	147,500
				$\Sigma p_0 q_0$	$\Sigma p_1 q_0$

Using the Laspeyre formula, Index $= \dfrac{147,500}{95,000} \times 100 = 155$

(2) *Paasche price index:*

Item	p_0	p_1	q_1	$p_0 \times q_1$ (£)	$p_1 \times q_1$ (£)
Bread	45p	50p	100,000	45,000	50,000
Cheese	500p	1,000p	15,000	75,000	150,000
Ale	900p	750p	3,000	27,000	22,500
				147,000	222,500
				$\Sigma p_0 q_1$	$\Sigma p_1 q_1$

Using the Paasche formula, Index $= \dfrac{222,500}{147,000} \times 100 = 151$

13. Laspeyre and Paasche indices contrasted. At this point the reader may well ask, what difference does it make which index is chosen? As regards the final figure, there will probably be very little difference unless there has been a substantial change in the purchasing pattern (in which case the index number will not tell us much anyway – after all, how many people are concerned to know by how much quill pens have gone up since Waterloo?). There are, however, two important practical points involving the computation and use of these indices.

(*a*) Paasche indices require actual quantities to be ascertained for *each* year of the series. This can be a big requirement. In contrast, a Laspeyre index requires quantities to be found for the base year only.

(*b*) With Paasche indices the denominator of the formula, $\Sigma p_0 q_1$, needs recomputing *every year* as q_1 changes yearly. In the case of Laspeyre numbers, however, the denominator, $\Sigma p_0 q_0$, always remains the same. Moreover, a consequence of this is that different years in a Laspeyre index can be directly compared with each other, whereas in a Paasche series the changing denominator means that different years can be compared *only* with the base year and not with each other.

For these reasons Laspeyre indices are much more common than Paasche indices.

14. The 'basket of goods' concept. When presenting index numbers statisticians sometimes simplify index concepts by referring to the cost of a 'fixed basket of goods' over two or more years. This refers, in fact, to no more than a quantity weighting of prices – the statistician providing quantity weights by imagining a basket of never-changing contents, say 2 kg of sugar and 4 litres of milk. The numbers '2' and '4' are then used to weight the sugar and milk prices in each year and a price index calculated in the normal way.

Other indices

Although the above indices are the ones most often met in practice, there are other important ones.

15. Price relative. A *price relative* is simply the price of an item in one year relative to another year, again expressed with 100 as base. Symbolically, it is

$$\frac{p_1}{p_0} \times 100$$

Thus, bread in our earlier example, being 45p in 19–0 and 50p in 19–5, has a price relative of

$$\frac{50p}{45p} \times 100 = \underline{\underline{111}}$$

16. Weighted average of price relatives index. Since each price relative is, in effect, a little one-item index number (*see* **5**), a composite index number can be obtained by *averaging* all the price relatives of items in a series. Again, weighting is necessary to allow for item importance.

Example

Find the weighted average of price relatives index for the figures in Table 15A with weights of 10, 7 and 3 respectively:

Item	19–0 price	19–5 price	Price relative	Weight	Price relative × weight
Bread	45p	50p	111	10	1,110
Cheese	500p	1,000p	200	7	1,400
Ale	900p	750p	83	3	249
				20	2,759

$$\therefore \text{Index} = \frac{2,759}{20} = \underline{\underline{138}}$$

Note: Remember that in a weighted average you divide by the sum of the weights (*see* 10:**18**).

The formula, therefore, is:

$$\textit{Weighted average of price relatives index} = \frac{\sum \left(\frac{p_1}{p_0} \times 100 \times w \right)}{\sum w}$$

It should be noted, incidentally, that the spread of weights in this index is much smaller than that of the weighted aggregative index. On the face of it this suggests that the drop in price of ale would have a greater influence. Yet the index is higher than before! The reason is that a price relative is quite different from a price: small prices, for instance, can have large relatives if they are unstable. For this reason, indices using price relatives must not be quantity-weighted (though they can be value-weighted).

17. Chain index numbers. A *chain index* is simply an ordinary index in which each period in the series uses the *previous period as base*. For instance, a simple example of a one-item chain index, using some of the data given in **5**, is given in Table 15C.

Such an index shows whether the *rate* of change is rising (rising numbers), falling (falling numbers) or constant (constant numbers) as well as the *extent* of the change from year to year. In the example given, it can be seen that although there is a steady increase in the sales of television sets, the increase each year, in relation to the total sales of the previous year, is on the whole falling.

Table 15C   *One-item chain index – TV sets sold*

Year	TV sets sold	Chain index
19–3	12,912	
19–4	18,671	$\frac{18,671}{12,912} \times 100 = 145$
19–5	21,200	$\frac{21,200}{18,671} \times 100 = 114$
19–6	28,633	$\frac{28,633}{21,200} \times 100 = 135$
19–7	35,028	$\frac{35,028}{28,633} \times 100 = 122$
19–8	40,650	$\frac{40,650}{35,028} \times 100 = 116$
19–9	44,531	$\frac{44,531}{40,650} \times 100 = 110$

In a multi-item index, such as one measuring the cost of living, a chain index is useful in so far as new items can be introduced. For instance, if it is wished to introduce home computers into such an index this particular year, then data for this year and last year only is required. Had a normal, non-chain index been used based on (say) 1975, there would probably have been no appropriate data available for that year and so it would be impossible to introduce home computers into the index.

18. Quantity indices. Apart from the indices of television set sales in **5** and Table 15C every index examined in this chapter has been a *price index*, i.e. a measure of price changes. There are other kinds of indices. An obvious one is an index of quantity.

A *quantity index* is one that measures changes in quantities. There are virtually as many different methods of computing a quantity index as there are a price index; in fact, formulae for quantity indices can be derived from those for price indices by simply interchanging the p and q symbols. Thus, a Laspeyre price index with a formula of

$$\frac{\Sigma p_1 q_0}{\Sigma p_0 q_0} \times 100$$

would become a Laspeyre *quantity* index with a formula of

$$\frac{\Sigma q_1 p_0}{\Sigma q_0 p_0} \times 100$$

Weighting, incidentally, is particularly necessary when constructing a multi-item quantity index as it is otherwise impossible to add together kilogrammes, litres, pairs, etc.

19. Value indices. Another group of indices relate to *value*, value being, of course, $p \times q$. Thus

$$\frac{\Sigma p_1 q_1}{\Sigma p_0 q_0} \times 100$$

is a *value index* since it compares values in the base year with values in a subsequent year.

Techniques involving index numbers

In this section we look at three techniques which employ index numbers. These are base changing, asset revaluation and average annual percentage changes.

20. Base changing. It sometimes happens that the user of an index wishes to change the base year. This often happens when two different series are to be compared, since it is unlikely that both will have the same base year, and so direct comparison between them would be difficult.

Example

Assume that an index of new television licences for the years 19–3 to 19–9, with 19–0 as base year, ran as follows:

Year:	19–3	19–4	19–5	19–6	19–7	19–8	19–9
Index:	210	230	250	300	360	410	500

Direct comparison with the index of television set sales computed in **5** is hardly possible. To obtain such a direct comparison it is necessary to change the base year of one of the series so that both have the same base.

21. Procedure for changing the base. The procedure for changing the base of a given series is as follows:

(a) Look up the series index number that relates to the year which is now to be the base year.

(b) Divide this number into each index number in the series and multiply by 100.

This will give a new series of index numbers with the new year as its base.

Note: Changing the base of a weighted index gives a series slightly different from that which would be obtained if the index had been computed entirely afresh with the new year as base. But for practical purposes the difference is rarely significant.

In the above example this means that to change the index of new television licences from a base year of 19–0 to a base of 19–5 all the figures will have to be divided by 250 (the 19–5 index number). To illustrate this, all the data relating to this example may be tabulated:

Year	TV sets sold (19–5 = 100) (see **5**)	TV licences (19–0 = 100)	TV licences (19–5 = 100)
19–3	61	210	$\dfrac{210}{250} \times 100 = 84$
19–4	88	230	$\dfrac{230}{250} \times 100 = 92$
19–5	**100**	250	$\dfrac{250}{250} \times 100 = \mathbf{100}$
19–6	135	300	$\dfrac{300}{250} \times 100 = 120$
19–7	165	360	$\dfrac{360}{250} \times 100 = 144$
19–8	192	410	$\dfrac{410}{250} \times 100 = 164$
19–9	210	500	$\dfrac{500}{250} \times 100 = 200$

Comparison of the two series is now possible and it indicates that while the rate of increase in the sales of television sets was initially greater than the rate of increase in licences, the trend was reversed in the later part of the series.

22. Index numbers and asset revaluation. These days, a very common application of index numbers involves the revaluation of assets. Clearly, because of inflation an asset bought in one year for a given figure would normally cost very much more if bought some years later. To adjust for inflation and convert the value of an asset in one year into the value of that same asset in another year the following formula can be employed:

$$\text{Adjusted value} = \text{Original value} \times \frac{\text{Price index for new year}}{\text{Price index for original year}}$$

where the price index selected is regarded as an acceptable measure of inflation.

Example

An asset was bought by a company for £72,000 in 19–3 when the retail price index (which management regard as an acceptable measure of inflation) stood at 157. Revalue the asset in 19–8 when the index had reached 296 as a result of inflation.

Solution

$$\text{Value in 19–8} = £72,000 \times \frac{296}{157} = \underline{\underline{£135,750}}$$

Note: This formula merely adjusts for inflation and does not allow for depreciation or any other non-inflationary factors which may affect the value of the asset.

23. Index numbers and percentage annual changes. Apart from the chain index, all types of index numbers show change relative to the base year. Sometimes, however, we need to know the percentage change from year to year and where this is so it is necessary to compute this percentage from the index numbers. Thus, we may want to know by what percentage TV set sales increased between 19–8 and 19–9 in the index shown in **21**. Finding this is quite straightforward. Since the index rose from 192 to 210 we can see that the increase is simply $(210-192)/192 = 9.375$ per cent. This type of computation can be shown by the following formula:

$$\text{Percentage change} = \frac{\text{later year index} - \text{earlier year index}}{\text{earlier year index}} \times 100$$

24. Index numbers and average annual percentage changes. The problem becomes a little more complicated when it is necessary to know the *average* annual percentage change over two or more years. For instance, if an index increases from 100 to 160 in three years the average annual percentage increase is *not* the total percentage increase (60) divided by 3 – i.e. it is not 20 per cent. If the annual percentage increase were 20 per cent, then while the index would certainly rise from 100 to 120 in the first year, in the second year it would rise from 120 to $120 + 20\%$ of $120 = 144$, and in the third year from 144 to $144 + 20\%$ of $144 = 172.8$ – which is certainly not 160.

Now, as it happens an average annual percentage change is a kind of geometric mean (*see* 10:**20**). And to find this mean, we use the formula:

$$\text{Average annual ratio change} = \sqrt[n]{\text{total ratio change}}$$

(i.e. the nth root of the ratio change) where: n is the number of years over which a total change occurs and a *ratio change* is the ratio of the later figure to an earlier, i.e. later figure/earlier figure.

So in our illustrative example the total ratio change is $160/100 = 1.6$. ∴ average annual ratio change $= \sqrt[3]{1.6} = 1.17$, and an annual change from 1 to 1.17 – i.e. an increase of 0.17 – is a percentage change of $+17$ (that 17 per cent is the correct average annual increase can be shown as follows: 100 in the first year becomes $100 + 17\%$ of $100 = 117$. 117 in the second year becomes $117 + 17\%$ of $117 = 136.9$. 136.9 in the third year becomes $136.9 + 17\%$ of $136.9 = 160.2$ – i.e. allowing for rounding, 160).

While the formula above is simple to remember, the reader who wants a more comprehensive formula should use:

Average annual percentage change
$$= 100 \, (\sqrt[n]{\text{later figure/earlier figure}} - 1)$$

Example

If an index stands at 215 one year and at 174.15 two years later, what is the average annual percentage change?

Average annual percentage change $= 100 \, (\sqrt[2]{174.15/215} - 1)$
$$= 100 \, (\sqrt[2]{.81} - 1)$$
$$= 100(.9 - 1) = -10\%$$

(Proof: 215 less $10\% = 193.5$. 193.5 less $10\% = 174.15$)

Index construction

So far we have looked at the definitions and mathematics of index numbers. Now we turn to the actual construction of such numbers and we will see that in such a construction four factors need to be considered:

(*a*) the purpose of the index;
(*b*) the selection of the items;
(*c*) the choice of weights;
(*d*) the choice of a base year.

25. Purpose of the index. The purpose of an index must be very carefully determined, for decisions relating to the other three factors will depend on the purpose. Moreover, the *interpretation* of the index will also depend on its purpose.

For example, an index constructed to measure change in building costs must not be used for revaluing machinery, or even the commercial value of a building, since such an index would not take into account changes in the values of land on which such buildings were situated.

26. Selecting the items. This can be the most difficult problem of all. Take the construction of a cost of living index. Obviously, bread should be included, but what about table wines? Heating costs ought to be included, but how about dishwashing machines? If home rentals are selected, should holiday rentals be selected too?

In the case of an index measuring employment, are part-time workers to be included? What about self-employed workers? In an export index, what should we do about imports which are immediately re-exported? Or imports returned for some reason to the overseas supplier? Whose share prices do we use for a share price index?

Moreover, the problem may arise as to *which* figures to take. Is a cost of living index to be based on prices in London or Manchester? Or in Little-Comely-on-the-Ouse?

The answers to such problems lie in defining the purpose of the index carefully and then selecting the items that will best achieve that purpose, although it must be realised there will always be differences of opinion.

Note that as regards the individual items selected:

(a) *They must be unambiguous.* An index of mortality from a given disease would be seriously distorted if improved diagnosis is attributing more and more deaths to the disease that previously had been attributed to other causes.

(b) *The values must be ascertainable.* The construction of an index relating to undetected murders would run into obvious difficulties.

27. Choice of weights. The problem here is to find weights which will result in each item being given its appropriate importance. Actual quantities may often be good weights but they are not invariably appropriate. If it were decided to take the quantities used by a typical household as weights, there would be some difficulty in determining a typical household. After all, people on the old age pension are very much concerned with cost of living figures and they would hardly be impressed with a heavy weighting for, say, private car travel.

But there is one factor which makes the problem of choosing weights easier. This is that a difference of opinion as regards weights does not, oddly enough, affect the index as much as one might expect. For instance, a completely revised weight of 200 for bread in Table 15B, i.e. a doubling of the original weight (which was already over *nine* times as big as the other weights combined), results in an index of 139 as against a previous number of 151 – a mere change of 8 per cent. Smaller revisions would result in smaller differences and this indicates that hair-splitting as regards weights is rarely worthwhile.

28. Choice of a base year. Generally speaking, the year chosen as base should be:

(*a*) a reasonably normal year; and

(*b*) not too distant.

Sometimes a year which is significant within the series may be chosen; for example, the year a Commonwealth country attained its independence might be an appropriate base year for an index of that country's production, or the year of privatisation might make a logical base year for the performance of a previously nationalised enterprise.

Choosing a freak year is a favourite trick of those who use statistics to mislead. A dishonest capitalist could choose a record year for profits as base and so 'prove' subsequent profits to be pitifully low. A dishonest trade unionist could similarly choose a year of exceptionally full employment to 'prove' that current unemployment is intolerably high. And a dishonest politician could use the year before privatisation to prove the virtues of privatisation when he had carefully massaged the figures of the years just prior to that event.

29. Which index should be used? To conclude, the bewildered reader may ask 'Which index should I use?' To this question there can only be one answer: *it depends on the circumstances*. The Laspeyre index is a good all-round index but it cannot be used if weighting figures (normally quantity) are unobtainable. In that case, the weighted average of price relatives, with weights determined on some other basis, may be appropriate. The purpose of the index is highly relevant. A chain index is obviously called for when the purpose is to indicate to what extent figures have changed in relation to the previous year. A chain index is valuable, too, where new items may need to be added and old items removed.

Finally, it must be emphasised once more that an index number is only an attempt to summarise a whole mass of data in one figure. Such a figure must inevitably be subject to many limitations and it is the responsibility of the user to balance all factors and judge:

(*a*) which type of index is appropriate; and
(*b*) the *real* significance of any single index number in the series.

Progress test 15

(*Answers in Appendix 4*)

1. From the data below, and using 19–3 as base where appropriate, prepare:

(*i*) a Laspeyre price index;
(*ii*) a Paasche price index;
(*iii*) a weighted average of price relatives, using the weighting: A, 5; B, 3; C, 2.

	19–3		19–4		19–5		19–6	
Item	Price £	Quan-tity	Price £	Quan-tity	Price £	Quan-tity	Price £	Quan-tity
A	0.20	20	0.25	24	0.35	20	0.50	18
B	0.25	12	0.25	16	0.10	20	0.12½	16
C	1.00	3	2.00	2	2.00	3	2.00	4

2. An examination of the last five years of the plant register of a company (which has the policy of depreciating its plant at 15 per cent per year, straight line – plant bought during the year taking a full 15 per cent for the year in which it was bought) revealed the following figures:

Year	Plant bought during year	Depreciation charged for year	Relevant mid-year price index
19–5	£35,000	£5,250	184
19–6	42,000	11,550	201
19–7	88,000	24,750	212
19–8	51,000	32,400	234
19–9	82,000	44,700	256
Total	£298,000	£118,650	

You are required to:

(a) revalue the plant cost at current prices at the end of 19–9 (when the index stood at 268) on the assumption that all plant was bought mid-year;

(b) compute the additional amount of depreciation that must be charged in order to update the aggregate depreciation charge so that the total aggregate depreciation reflects the company's depreciation policy.

3. If the sales index shown in **21** relates to one company's sales, suggest possible reasons why its rate of sales increase slowed down relative to the increase in TV licences.

4. A company calculates an index at the end of each period of six months to give an impression of the increase in raw material prices. The index was constructed so that its value on 31 December 1976 was 100. The values of the index over a three-year period were as follows:

6 months ending	Index
30 June 1977	105
31 December 1977	114
30 June 1978	123
31 December 1978	130
30 June 1979	139
31 December 1979	149

Required:

(*a*)(*i*) Determine the 6-month period in which the greatest percentage increase occurred in raw material prices. Assume for simplicity that all 6-month periods contain the same number of days.

(*ii*) Show that the mean annual percentage increase in raw material prices during the three year period was just over 14 per cent and calculate the value of the index on 31 December 1981 if this mean rate of increase is maintained.

(*b*) Indices calculated by the company for wage rates and for variable overheads per unit, again using 31 December 1976 as a base, gave values of 163 and 127, respectively, for 31 December 1979. A certain class of product has a variable cost which comprises 40 per cent raw material, 40 per cent labour and 20 per cent variable overheads. Estimate the mean annual percentage increase in variable cost of this class of product over the three year period, 31 December 1976 to 31 December 1979.

(*ACCA*)

Assignment

From the data in question 1 compute:

(*a*) for each item:
  (*i*) a quantity index;
  (*ii*) a price index;
(*b*)(*i*) a Laspeyre quantity index;
  (*ii*) a Paasche quantity index;
(*c*) a value index.

16
Statistical sources

So far we have looked almost exclusively at how you can compile your own statistics. But there is already a wealth of compiled statistics available for your immediate use. Moreover, such statistics are most unlikely to be compilable by you, yourself, since they relate to the nation, or region, as a whole. Many of these statistics are of particular value to businesses since they indicate the potential opportunities for enterprise – as was said earlier, if you sell baths it is useful to know how many houses are planned to be built (*see* 12:**2**). So it is important that businesses know what and where the available information is.

In this chapter we will look briefly at these *statistical sources*. It should be said at once, however, that it is essential that any serious student of statistics must personally visit a major reference library and see for himself not only what statistics are available but the manner and form in which they are presented to the user. Only in this way can he obtain a feel of the volume, scope and organisation of the statistical data available.

Retail price index (RPI)

Probably the most quoted published statistic is the retail price index (RPI). It will pay us, then, to look at this particular index before we turn to the various statistical sources, both because of its widespread reproduction in many statistical publications and also as an example of how government statistics are compiled.

1. Computation of RPI. The first problem of compiling an index intended to measure retail prices is to determine what commodities should enter the index and to what extent. This is solved by taking a national survey of households to discover just what households purchase and to what extent. Because their expenditure patterns differ markedly from the normal, households with either limited means (e.g. pensioners) or high incomes (some 4 per cent of the total) are excluded – recognising, in fact, for this particular situation the perennial problem of extreme values that we discussed in 3:**10**. The households to be surveyed are then interviewed to find their patterns of expenditure (*see* **13**). And from this the items required for the RPI are selected and their relative importance assessed, this selection and assessment process resulting in a 'basket' of 350 goods and services grouped into 95 sections which are further collected into 11 weighting groups.

The index items selected, the next step is to find the retail prices. To do this a selection of towns, and shops within the towns, is made, the selection aiming at being representative of the towns nationwide and shops normally visited by the public. Then, on a predetermined Tuesday near the middle of each month, the actual prices in these shops are ascertained. Of course, this procedure only covers those of the 350 items which are sold in shops, and cannot include items such as electricity, mortgage rates and other services. In most instances the prices of these latter items can easily be obtained from national or regional offices and give no problems of ascertainment.

At the end of the exercise the statisticians find they have some 150,000 prices quotations. These are next built into the index in a manner best judged to reflect the retail prices across the country as a whole. Finally, the index figure itself is computed as a weighted average of price relatives index figure (*see* 15:**16**), i.e. by applying the predetermined weights to the price relatives and then relating the composite figure to the base year.

2. A typical compilation problem. As can be imagined, compiling the RPI is quite an extensive operation – and one that, as far as ascertaining prices goes, must be conducted monthly. Yet even the procedure above does not indicate how the various complications that arise in practice impose additional burdens on the operation. For instance, as prices rise and fall people switch the products they buy, buying more of the cheaper and less of the dearer. So the basket of goods and services changes a little, and generally the index must reflect this (e.g. if the basket had included a packet of Bloggins' pottage and the public switched to buying Fred's pottage then the basket would be amended). However, since the RPI is not intended to be seasonally adjusted, switching where prices are seasonal (e.g. as in the case of fruit and vegetables) is ignored since to allow for it would, in fact, result in ironing-out (i.e. seasonally adjusting) price changes.

3. RPI not a cost of living index. An important point to note is that the RPI is *not* a cost of living index. Indeed, the point in the previous paragraph illustrates this since, if it is possible to switch your purchasing according to seasonal prices, you can keep your cost of living relatively stable compared to the actual change in prices. And this is why the RPI is not intended to be seasonally adjusted. Again, a cost of living index would measure what you needed to buy to maintain a given living standard whereas the RPI measures the prices of what households actually buy.

4. The index for pensioners. Finally, it should be noted that although the expenditure of people with limited means is excluded from the RPI, they are not forgotten but have their own index published in the *Employment Gazette*.

Statistical publications

Without doubt, the one segment of society to whom national statistics are absolutely vital is the state (as the word 'statistics' implies). Without up-to-date statistics sound economic government is virtually impossible. It is not surprising, then, that the great majority of sources of general statistical data are government publications and it is these we look at in this section.

5. Guide to Official Statistics. The major problem with statistical sources in this day and age is that there are so many of them. As a result the hardest part of making use of published statistics is knowing where to look. To help with this problem the Central Statistical Office publishes a *Guide to Official Statistics*. This Guide essentially lists all the official, and some non-official, sources of data for the UK, Isle of Man and Channel Islands.

For somebody looking for a particular set of statistics for the first time, the best place to start searching is the 'Contents' page of the Guide. From this it should be possible to pick out the broad category of the inquirer's interest (e.g. housing, exchange rates), and the page in the Guide on which the sources are detailed. This latter page not only lists the relevant sources but also gives some indication of the contents of each publication.

6. Monthly Digest of Statistics. Probably the most frequently used statistical source is the *Monthly Digest of Statistics*. As indicated, this is a monthly publication, and it is available relatively soon after the end of the month to which it relates. It is, then, one of the most up-to-date statistical publications though this sometimes means that some figures are later subject to revision. The contents include statistics on: national income and expenditure; population and vital statistics; employment; social services; external trade; overseas and home finance; prices and wages; the retail price index; and a wide variety of data related to all economic areas (e.g. agriculture and food, energy, chemicals, engineering, textiles, construction, transport and retailing). The data is presented primarily in the form of tables showing data not only for the month of issue but for many months previously, and also totals for even earlier years.

7. Annual Abstract of Statistics. The *Annual Abstract of Statistics* is essentially a considerably extended version of the *Monthly Digest*. It is, however, only published annually (and hence the figures, too, are only yearly totals) and then not until some time after the end of the year.

8. Business Monitor, PQ Series. *Business Monitor, PQ Series,* is published quarterly and is, perhaps, the most generally useful statistical source after the *Monthly Digest*. It is up to date (being published some four months after the end of the quarter) and records for some seven years the annual sales, exports and imports for a wide and detailed variety of product groups and activities. It is published as a large number of slim booklets (5–20 pages long), each covering a major product or activity group – which means a particular enterprise need only purchase the booklet for the product of its interest. Within each booklet statistical details are given for a very large number of product subgroups (e.g. plastic vending cups, gummed labels, kitchen knives) and this makes the publication of considerable value to businesses engaged on work in these specific and often finely differentiated products.

9. Business Monitor, PA Series. *Business Monitor* is also published as a *PA Series*. These form an annual publication and replace the old *Census of Production*. Again, each issue comprises a large number of slim booklets, each covering an aspect of business and each detailing a variety of statistics relating to production and production facilities.

10. Economic Trends. *Economic Trends* is published monthly and, indeed, is rather akin to the *Monthly Digest* – at least, as far as the economic statistics are concerned. It is, however, more concerned with national *topics* (e.g. stock changes, balance of payments current account, UK bank lending to UK residents, financial transactions of the public sector) than industry groupings, although there is, of course, overlap between the two publications.

In addition to a wide variety of statistical tables, issues include articles on selected aspects of statistics (e.g. effect of taxes and benefits on household incomes) and also a number of graphs (including some semi-log graphs).

11. Employment Gazette. The *Employment Gazette* is another monthly publication that, understandably, gives employment statistics. Issues are split between articles on employment (including general news topics) and tables of statistics – the latter including data on: employment (numbers, overtime and short-time, hours of work); unemployment (analysed by groups, e.g. regions, age, duration) with international comparisons; vacancies; industrial disputes; earnings; retail prices (including the special RPI figures for people with limited means – *see* **4**); household spending; tourism.

12. New Earnings Survey. The *New Earnings Survey* is an annual publication which covers all earnings statistics in Great Britain. It is based on a survey in respect of a 1 per cent random sample of all employees who are members of PAYE schemes – although the data is, in fact, obtained from employers by means of a postal questionnaire. The statistics include earnings and hours for a variety of industries, occupations, regions and age groups – the data being cross-analysed in virtually every viable way. Also included are statistics relating to national wage agreements.

13. Family Expenditure Survey. The *Family Expenditure Survey* aims to provide economic statistics relating to households in Great Britain. The figures are based on extensive interviews with a random sample of households. In the tables the data is classified by types of households, e.g. region, number of adults, number of children, level of household income, age of household head, form of household tenure, and employment status (e.g. retired, self-employed).

14. General Household Survey. Not to be confused with the *Family Expenditure Survey* is the *General Household Survey*. This publication, while certainly looking at households, primarily relates to *people* in households, e.g. it publishes statistics covering marriage, employment, education, health and leisure.

15. Overseas statistics. For overseas statistics probably the best general source is the *World Development Report* which is published annually by the World Bank. Although much more concerned with economic statistics than perhaps commercial statistics, it would appear to be one of the best introductions to overseas statistics. The Report is not overburdened with figures – indeed, informative articles, liberally illustrated and attractively presented, constitute the bulk of the book – but the tables produced allow a comprehensive overview of the 120-odd countries (from Albania to Zimbabwe) to which the statistics relate.

On-line database statistics

Valuable as the current extensive range of statistical publications is, there can be little doubt that as statistical sources they will ultimately – and arguably very soon – be relegated to a secondary role. This change of emphasis is due to the increasing viability of using on-line databases.

16. On-line databases. Statistics might have been invented for the database which is essentially no more than the organised collection of data held in a computer. Computers, of course, can 'memorise', organise and retrieve input data quickly, easily and cheaply and where masses of data call for this form of treatment the computer is the natural tool. It is, though, one thing for data to be held on a computer but quite another for that data to be available to someone perhaps two or three hundred miles away. But with modern telecommunications even this problem is beginning to disappear and it is becoming increasingly easy to call up the database computer from a distant terminal and for the terminal user to obtain just whatever data (statistics) he requires. In other words, the user can in essence have flashed up on his screen the kind of statistical data that he would otherwise find on the printed page. And this form of information is termed *on-line database statistics*.

17. Information available. It is probably true to say that there are hardly any general statistics which are obtainable from published journals and abstracts which cannot also be obtained from on-line databases – and the coverage is growing. At the moment most such information is being provided by private organisations and anyone wishing to use this form of statistical reporting needs to consult a database sourcebook. Typical, perhaps, of the kinds of surveys handled is the *Target Group Index* which compiles data each year from 24,000 adults in the UK, the data covering such things as products bought and used, media interest (e.g. papers read, TV viewing) and demographic information (e.g. age, marital status).

18. The economics of on-line database statistics. This new form of statistical sources completely reverses the economics of data distribution (though the economics of data compilation remain unchanged). Printed publications involve a high fixed cost of production and distribution to libraries, and thereafter a low variable cost per item of data actually perused. With databases the fixed cost of organising the database is relatively very low but the variable perusal cost is high in respect of, collectively, the communication costs, the terminal cost and, to a much smaller extent, the database computer accessing cost.

But these new economic impacts are very much for the better. With printed publications there is massive waste. By far the vast majority of figures in a given copy of an issue are never even glanced at – the publications moulder silently on the library shelves. Indeed, it is possible that many figures are never ever used anywhere which means that the typesetting, printing, paper, proofreading, and distribution costs in respect of these figures is totally wasted. But, of course, since no one ever knows what figures will be wanted, they all have to be included in the publications. In the case of databases, however, figures are only drawn from the database *if they are wanted*, so costs are only incurred in respect of wanted data. Of course, data may be unnecessarily *compiled* and stored in a database. This cost, then, cannot be avoided, but at least the reproduction and distribution of unwanted data is eliminated.

A further economic benefit of this new approach is that the cost of providing statistics is more directly chargeable to the beneficiary of the data. He who wants must now pay the higher variable costs of access himself. In the case of printed publications, people with a low requirement for statistics carry an undue proportion of the cost – either by having to purchase a publication with far, far more figures than they actually require or by subscribing via taxes for libraries to purchase the material they never need. And if all this were not enough, yet another additional benefit arises from the fact that the database operator can monitor which data is actually accessed – i.e. wanted – and this means that subsequent compilations of statistics can concentrate on the more relevant data with only a relatively brief glance, as it were, at the rarely wanted information.

19. Advantages of on-line databases to the user. Notwith-standing the general economic benefits of database statistics, this type of source has specific benefits for the user, these including:

(a) *Much faster search times.* Even if the user has his own copy of a printed publication it is probable that he can find a particular set of figures more quickly using the database than thumbing through pages. And if he has not his own copy of the publication then it means a lengthy trip to the library.

(b) *More up-to-date statistics.* With printed statistics there is a long period during which the data is typeset, the proofs produced, read and corrected, the pages printed and a copy of the publication distributed, catalogued and placed on library shelves. This time period is virtually wholly eliminated with database statistics since the time to input statistics into a database is relatively insignificant.

(c) *More detailed statistics.* Far more figures are compiled than are ever published in printed statistics since the fixed cost per figure printed makes the reproduction of the more detailed data uneconomic (and also, of course, library shelving is too limited to store the massive publications which would result from publishing all the figures). This, however, is not so in the case of the database. Although additional items do add a little to the cost of the database, this is rarely excessive per unit of data (though there is a limit even with databases). So much more detailed statistics are potentially available to the user.

(d) *Personalised print-outs.* The use of on-line databases enables the user to have all the statistical data he personally wants printed out quickly and accurately in his own format (e.g. one sheet of paper can display figures relating to any variety of statistical data on a given subject). And, of course, if the user has not his own printed copy of a statistical publication, a terminal print-out saves him the bother of obtaining a series of photocopies. Moreover, if the supplier's computer is large enough to handle the likely demands upon it, it may also be possible to abstract statistics in the form of personal statistical arrangement, e.g. have printed out statistics in respect of the leisure activities of pet-owning, single-parent, flat-dwelling families who normally holiday on the Costa Brava.

20. Disadvantages of on-line databases to the user. On-line databases, however, do have disadvantages of which the following are probably of greatest importance:

(a) *Not possible to browse.* One of the advantages of the printed page of statistics is that you can let your eye wander over both the classifications and the individual figures. This browsing can often alert you to facts which you never suspected would be of relevance to your inquiry. Indeed, a modern theory of progress asserts that a very great deal of progress is the result of chance discoveries. With a database it is difficult (though not impossible) to browse since the computer can only transmit to you the data you ask for and not any data you do not. So your opportunity for chance discoveries is very much reduced.

(b) *Currently, little standardisation of procedure.* At the moment the fact that the technology is new and that private organisations are the main suppliers of database statistics means that there is little standardisation of the accessing procedure. This in turn means that users have to spend valuable time familiarising themselves with procedures on each occasion they wish to access a new database.

(*c*) *No friendly librarian.* There can be little doubt that when you are looking for unfamiliar statistics there is no substitute for the friendly librarian. Such a person not only knows where you can find the data you want but also can alert you to other statistics which may have some, or even more, relevance to your investigation. Additionally, he or she will probably be able to produce supporting material detailing how the statistics you are using were compiled (which helps considerably in evaluating your material), and possibly articles in journals which discuss the compilation procedures or even criticise the quality of the resulting figures.

21. Economic benefits to the user. The economics of the user employing on-line databases was mentioned neither among the advantages nor the disadvantages of this form of statistical source. The fact is that it may or may not be cheaper than utilising printed publications. Clearly, the user will have two major costs – the telecommunication cost and the access cost charged by the data supplier. Additionally, to the extent that his terminal is used exclusively or primarily to access the databases, there are his hardware and software costs. Against these can be set either the purchase costs of the statistical publications or the cost of visiting a reference library. Oddly enough the extent of his need for statistical data does not affect the economic balance much – the very occasional user may find that it is cheaper to visit the public library once in a while than pay the subscription cost to a database operator, while the very frequent user may find that it is cheaper to buy the printed publications – even though he needs only a fraction of the data he obtains – than pay a large number of telecommunication and access charges.

But, of course, the decision is not just economic – the quality of the information is also highly relevant. And, as things are now, this is almost certainly the touchstone of the decision to use on-line databases since the current charges for the facility are very high indeed – almost certainly prohibitively high for the normal run-of-the-mill statistics. Nevertheless, with falling telecommunication and computer costs and the growing expansion of database statistics there can be no doubt that the on-line database is here to stay – and ultimately to dominate the statistical sources domain.

Progress test 16

1. How is the RPI compiled? (**1–4**)

2. Where is the best place to start looking for an unfamiliar series of statistics? (**5**)

3. Distinguish between:

(*a*) the PQ and the PA series of *Business Monitor* (**8,9**);

(*b*) the *Family Expenditure Survey* and the *General Household Survey* (**13,14**).

4. State the advantages and disadvantages of on-line database statistics to the user (**19,20**).

5. Describe the economic aspects of on-line database statistics (**18,21**).

Assignments

1. Find three statistical sources not mentioned in this chapter which you feel provide useful statistical data in your area of study (e.g. business, law, transport, etc.) and write a report explaining the reasons for your choice.

2. On the assumption that you are the manager of a business in the heavy vehicle industry, ascertain what statistics, and their sources, are available to you which you believe you would find helpful in your work. What other statistics, which are not available, would you like to see and where or how might you expect these to be provided?

17
Time series

Our next topic is time series.

It should be made clear at once that we are now entering a controversial field. There are some statisticians who argue that a time series analysis has very little value (e.g. Moroney, *Facts from figures*). The reason for this is that such analyses are used almost exclusively to make *forecasts*. But statistics is not about what will be, but what is. Look at the rest of the book and you will see this is so. True, our regression analyses allowed us to 'forecast' what the value of one variable would be, given the value of another, but such forecasting was merely computing a value from an established, and often causal, relationship. It is one thing to forecast that next year there will be an eclipse of the moon, but quite another that next year there will be *x* American tourists visiting Europe. It only needs a bomb in a boulevard or a Chernobyl to rubbish the figures entirely.

A spin-off from this situation is what could be an illustration of the principle that while there is probably only one right way to do a thing, there are dozens of wrong ways. In this chapter you will meet alternative ways of computing time series values – each giving a different answer – just as once there were alternative ways of computing how many angels could sit on the head of a pin.

But this is neither the time nor the place to enter into debate. You need the conventional wisdom to pass your examinations. From here on it will be presented to you – without comment.

Introduction

As we saw in 1:5 a time series is a sequence of variable values that change over time. Very often there is a pattern underlying such figures and the object of a time series analysis is to tease out this pattern. In order to illustrate the methods which will be described we will use the following time series data:

Table 17A *Time series*
Fred's quarry sales 19–3 to 19–6 (tonnes)

Year	Quarter 1	Quarter 2	Quarter 3	Quarter 4	Total
19–3	672	636	680	704	2692
19–4	744	700	756	784	2984
19–5	828	800	840	880	3348
19–6	936	860	944	972	3712

Note, however, that four years is really too short a time to compute reliable values and that in practice a longer period should be used.

1. How not to read a time series. Looking at the figures above, can you say which is the busiest quarter? At first glance quarter 4 appears the busiest, but when we graph the figures (*see* Fig. 17.1) it becomes apparent that quarter 1 is the busiest. As will be seen, the illusion arises because quarter 4 benefits from a rising trend so that by

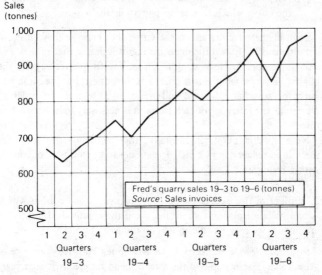

Figure 17.1 *Time series: Table 17A*

the *year-end* its sales are greater than quarter 1 of the same year. But it is not the busiest of all the quarters in all the years.

2. Factors influencing a time series. If a graph is drawn of a time series over a long enough span of time the following features may well be seen:

(*a*) *Seasonal variation* – this is a regular up-and-down pattern that repeats *annually* and is due to the effect of seasons on the variable.

(*b*) *Cyclical variation* – this is a regular up-and-down pattern that repeats over a span of years. In the main it reflects the boom/depression economic cycle.

Note: As the boom/depression cyclical variation is nowadays much more dependent on government policy (one's own or someone else's) than economic rhythms, there is rarely any underlying *time-based* pattern to be analysed. This form of variation is, then, ignored in the rest of this chapter.

(*c*) *Trend* – this is an overall tendency for the curve to rise (or fall).

(*d*) *Random (or residual) variations* – these are odd movements of the curve which fit into no pattern at all.

Each of these factors affects the curve and, because they all do so simultaneously, it is difficult to distinguish clearly the influence of any single one. As we saw, although the fourth quarter appears the busiest, in actual fact the first quarter is the busiest, but the combined effect of seasonal influence and trend disguise this. In order to determine the influence of each factor, it is necessary to isolate each in turn.

Note at this point, incidentally, that a time series graph is sometimes referred to as a *historigram* – not to be confused, of course, with a histogram.

3. Object of the exercise. There are two very good reasons why it is worthwhile isolating each factor, for if their individual influences are known we may be able to do two things.

(*a*) *Predict future values of the variable.* Such knowledge may be of great value. For example, if the total demand for electricity ten years from now can be estimated, the facilities that will be needed in ten years and which have a ten-year constructional period can be started at once, not in (say) six years' time, when it will be too late.

(*b*) *Control events.* In exercising control it is often very important to know at the earliest possible moment should a new element enter the situation. The interaction of existing factors tends to hide the appearance of a new element until it has already had unforeseen effects. Analysis of the series helps to reveal 'intruders' at an early stage. If, when the actual figures are received, they differ from the predicted ones by an amount greater than could be explained by random variation, there is a strong probability that a new influence has entered the series, altering either the trend or the seasonal pattern. In **10** an example is given where sales appear to be maintaining a continuous rise to the end of the year but in reality stopped rising six months earlier. It should be noted, incidentally, that in this application we are concerned with what is and not what will be.

Seasonal variation

There are several ways of analysing a time series to isolate the seasonal variation, but probably the most generally satisfactory ones utilise the *method of moving averages*, which involve computing moving averages (*see* 5:**20**). Each average, of course, eliminates seasonal influence and is located at the centre of the period to which it relates. The actual figure for the season it is located against is then used to measure the extent to which seasonal factors cause the actual to deviate from the moving average.

4. Method of moving averages. Any analytical technique that uses moving averages to separate out seasonal factors is known as a *method of moving averages*. In this section we will only look at techniques that adopt this method.

It should, perhaps, be appreciated that a moving average only eliminates seasonal variations from a series if the span of the seasons covered by the average is exactly a seasonal cycle (e.g. if the series is one of quarterly figures, then the average has to be one of four quarters – finding the moving average of five quarters does nothing to produce a moving average free of seasonal variations).

Table 17B *Analysis of seasonal variation in Table 17A*

Year	Quarter	Actual sales	MAT	Moving average	Centred average†	Seasonal variation Mult‡	Add††
19–3	1	672				§	§
	2	636				§	§
			2,692*	673			
	3	680			682	100	−2
			2,764	691			
	4	704			699	101	+5
			2,828	707			
19–4	1	744			716½	104	+27½
			2,904	726			
	2	700			736	95	−36
			2,984	746			
	3	756			756½	100	−½
			3,068	767			
	4	784			779½	101	+4½
			3,168	792			
19–5	1	828			802½	103	+25½
			3,252	813			
	2	800			825	97	−25
			3,348	837			
	3	840			850½	99	−10½
			3,456	864			
	4	880			871½	101	+8½
			3,516	879			
19–6	1	936			892	105	+44
			3,620	905			
	2	860			916½	94	−56½
			3,712	928			
	3	944				§	§
	4	972				§	§

†Centred average: mean of two moving averages on either side of it.
‡Multiplicative seasonal variation.

$$\frac{\text{Actual figure}}{\text{Centred average}} \times 100$$

(Be careful not to work with this formula upside down.)
††Additive seasonal variation.

Actual figure − Centred average

*This MAT figure was found by adding the sales of the four quarters of 19–3. Since this is only a step on the way to finding a moving average it is located at the midpoint of this period, i.e. between quarters 2 and 3.
§No figures possible for these quarters.

5. Centred averages. The idea of comparing an actual figure with the moving average for the same period is fine in theory, but unfortunately if there is an even number of seasons, the moving average will inevitably be centred between two seasons. To overcome this difficulty the moving averages on either side of a particular season are averaged to give a *centred average*, centred on that season. This will probably be better understood by studying the example used to illustrate the procedure given below.

6. Multiplicative and additive methods. There are two ways in which a difference between an actual figure and a moving average can be measured – by finding the actual as a percentage of the average or by finding the absolute difference between the two. If the former way is chosen then the *multiplicative method* of moving averages is being used. If the latter, the *additive method* of moving averages.

7. Method of moving averages: procedure. Whichever method is selected the steps in the procedure are the same except for step (*e*). The procedure, then, for both methods is as follows (*see* Table 17B which uses the data given in Table 17A).

(*a*) List the series vertically.

(*b*) Compute the moving totals and write these at the midpoint of the periods to which they relate.

(*c*) Compute the moving averages.

(*d*) If there is an even number of seasons, average the adjacent moving averages to give centred averages centred on each season.

(*e*)

(*i*) *Multiplicative method.* Compute (100 × actual figure/centred average) – *see* 'Multiplicative' column, Table 17B.

(*ii*) *Additive method.* Compute (actual figure – centred average) – see 'Additive' column, Table 17B.

This step gives the seasonal variations of each of the individual seasons. However, examining these will show that a given season has slightly different values at different points of the table – e.g. quarter 2 has values of 95, 97 and 94 (multiplicative) and -36, -25 and $-56\frac{1}{2}$ (additive). These inconsistencies need to be eliminated – which leads us to the next step.

(*f*) Find the means for each season of the individual seasonal variations. Averaging for the two methods gives us:

Year	Multiplicative				Additive			
	Q1	Q2	Q3	Q4	Q1	Q2	Q3	Q4
19–3	–	–	100	101	–	–	–2	+5
19–4	104	95	100	101	$+27\frac{1}{2}$	–36	$-\frac{1}{2}$	$+4\frac{1}{2}$
19–5	103	97	99	101	$+25\frac{1}{2}$	–25	$-10\frac{1}{2}$	$+8\frac{1}{2}$
19–6	105	94	–	–	+44	$-56\frac{1}{2}$	–	–
Total	312	286	299	303	+97	$-117\frac{1}{2}$	–13	+18
Average	104	$95\frac{1}{3}$	$99\frac{2}{3}$	101	$+32\frac{1}{3}$	–39	$-4\frac{1}{3}$	+6
Total		400				–5		

(*g*) Adjust by applying formula:

$$\text{Required adjustment} = \frac{\text{Required total} - \text{actual total}}{\text{No. of seasons}} \text{per season}$$

and applying this adjustment pro rata over the seasonal variations found in (*f*). Specifically, for each method:

(*i*) *Multiplicative method.* Clearly, if we are measuring the seasonal variation of a season from the non-seasonal average then since the average of all the seasonal variations in a year must be 100, the sum of the seasonal variations must be 100 × the number of seasons – e.g. the required total here is 100 × 4 = 400. So:

Required adjustment = (400 – 400)/4 = 0

(*ii*) *Additive method.* In this method the pluses must cancel out the minuses since the sum of the differences from a mean must be zero. So here:

Required adjustment = (0 – –5)/4 = $+1\frac{1}{4}$ per season

When an adjustment as small as this is needed it is probably better to ignore the principle of spreading the adjustment pro rata and instead spread it arbitrarily – using it more to eliminate fractions than achieve mathematical exactitudes. So in our illustrative case we would end with: achieve mathematical exactitudes. So in our illustrative case we would end with:

Method	Q1	Q2	Q3	Q4	Total
Multiplicative	104	95	100	101	400
Additive	+34	−38	−3	+7	0

It should be noted that this procedure can be applied to any kind of season. If months are used, the only effects are that there are more calculations and, in the case of the multiplicative method, that the adjustment in step (g) would entail making the sum of the seasonal variations 1,200. Note, too, that seasons need not be seasons of a year. Morning, afternoon and night are 'seasons' in relation to electricity demands, for instance; so are certain times of day for passenger transport services.

8. Theory underlying the method of moving averages. From the computations above, quarter 1 emerges as the busiest season, with seasonal variations of 104 and +34 respectively. Quarter 4 earlier appeared the busiest because, as we said, in this series there is a strong upwards trend: if there were no seasonal variations at all, the last quarter would be distinctly higher than the first simply because, coming later in time, it would benefit from the trend. In the series we have examined, the trend is so steep that although quarter 4 was not as busy seasonally as quarter 1, sales in the last quarter were always higher than in the first quarter of the same year.

It is in order to eliminate the distortion of the seasonal figures by the trend that the method of moving averages is used. The approach is to find out what the figure for each season would be if there were no seasonal variation, and then relate the actual figure to this as a percentage or an absolute difference. In 5:**16** we saw that moving values eliminate seasonal variations and therefore moving averages are used.

The adjustment in step (g) is needed because if some seasons are above average some must be below and, of course,

$$\frac{\Sigma(Seasonal\ variations)}{Number\ of\ seasons}$$

must equal 100 or zero. Accordingly, we adjust the figures so that this, in fact, becomes so.

9. Multiplicative or additive. At this juncture you may wonder which of the two methods of moving averages should be used. In this context you should appreciate that seasonal swings are normally pro rata to the magnitude of the figures – e.g. when a small business grows bigger the absolute difference between the seasonal figures grows but the relative percentage differences stay much the same. As can be seen from the additive part of Table 17B, the differences grew as the years progressed whereas in the multiplicative part of the table the percentages remain very much more stable. This, of course, was simply a reflection of the fact that the trend was rising rapidly.

It follows from this that the additive method should really only be used where the trend is relatively flat. Given that the multiplicative method is equally able to handle both flat and steep trends, it is clearly the better general purpose method, and if a choice is available should normally be selected.

De-seasonalising a time series

When examining figures subject to seasonal variation one of the problems is to know whether, say, a relatively high figure in a busy season is due wholly to seasonal factors or whether some other factors are involved. If the seasonal variations are known, they can be used to remove the seasonal influences from the figures. The resulting figures are then said to be *de-seasonalised* (or *seasonally adjusted*). The influence of factors other than the seasonal variation can then be seen.

10. Computation of de-seasonalised figures. To remove the seasonable influence in an actual seasonal figure, the following formulae are used:

(a) *Multiplicative method*:

De-seasonalised figure = 100 × *actual figure/seasonal variation*

(b) *Additive method*:

De-seasonalised figure = *actual figure – seasonal variation*

Example

Assume that when the 19–7 sales figures for Fred's quarry were received they were (in tonnes):
Quarter 1: 1020. Quarter 2: 960. Quarter 3: 1010. Quarter 4: 1020. De-seasonalise these figures and comment on your results.

		Multiplicative method		*Additive method*	
Qr	*Actual figure*	*Seasonal variation*	*De-seasonalised figure*	*Seasonal variation*	*De-seasonalised figure*
1	1020	104	$100 \times 1020/104 = 981$	$+34$	$1020 - +34 = 986$
2	960	95	$100 \times 960/95 = 1010$	-38	$960 - -38 = 998$
3	1010	100	$100 \times 1010/100 = 1010$	-3	$1010 - -3 = 1013$
4	1020	101	$100 \times 1020/101 = 1010$	$+7$	$1020 - +7 = 1013$

The actual figures suggest that the previously upward trend was reversed sharply in the second quarter but that in the third and fourth quarters sales reverted to their upward movement. However, the multiplicative method de-seasonalised figures indicate that the upward trend continued up to the second quarter and then levelled out, while the additive method de-seasonalised figures, though confirming the levelling out by the end of the year, indicate that the sales initially continued to rise through to the third quarter.

Trend

Now that we have seen how to isolate the seasonal variations we can examine the next factor, the *trend*.

Since moving averages indicate trend, it might be thought that the trend of any series is sufficiently given by the fifth colum of Table 17B. Unfortunately, moving averages also include random variations along with the trend, so they do not result in the straight trend lines which are really needed for prediction and control. Nevertheless, the series of moving averages is often referred to as the trend (*see* Fig. 17.2).

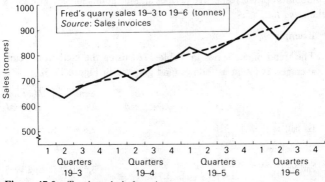

Figure 17.2 *Trend: method of moving averages*

In this section, though, we look at how a straight-line trend can be computed. First, note that there are two principal ways of finding such a trend:

(a) the method of semi-averages;
(b) the method of least squares.

Remember, incidentally, that when graphing any trend figures for the purpose of further mathematical work it is necessary always to plot totals at the *midpoint* of the period to which they refer (*see* 5:**14**(b)(ii)).

11. The method of semi-averages. This is by far the easier of the two methods of finding the trend, but it is rather crude and is apt to be inaccurate if there are any extreme values in the series. However, providing these limitations are borne in mind, it can be usefully employed in appropriate circumstances.

12. The method of semi-averages: procedure. The following procedure identifies the trend by the method of semi-averages.

(a) The annual totals are computed (or totals of complete cycles, should the seasons be other than of the normal yearly kind).
(b) The series is divided into two halves, each containing a complete number of years. If the overall series contains an odd number of years, the middle year is omitted.
(c) The mean value of each half is computed.
(d) These two mean values are then plotted on a graph at the midpoints of their respective periods and the points joined. This gives the *trend line*.

Example

The trend of the series in Table 17A, using the method of semi-averages, is found as follows (and graphed in Figure 17.3).

	Year	Total sales for year (tonnes)	Semi-average	Midpoint of period
1st half	19–3	2,692	2,838	End 19–3
	19–4	2,984		
Dividing line				
2nd half	19–5	3,348	3,530	End 19–5
	19–6	3,712		

13. Using the trend line for prediction. If the trend line is projected to the right, an estimate of future yearly totals can be read from it. It must be borne in mind, though, that on this particular type of graph, annual totals must be read at the midpoint of the year concerned. Thus, in Fig. 17.3 for example, the estimate of sales in 19–7 is 4,050 tonnes.

Sales
(tonnes)

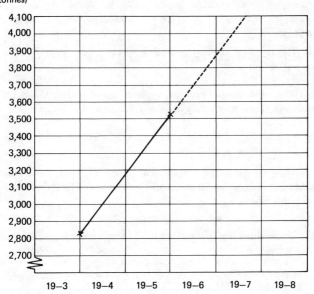

Figure 17.3 *Trend: method of semi-averages*

14. The method of least squares. Readers who studied Chapter 13 will remember the method of least squares. A moment's reflection will show that a trend line is, in fact, an ordinary regression line of the variable concerned on time (since we wish to predict variable value from time and not vice versa). This means that the reader who wishes to use the method of least squares to find a trend line may compute the regression line in the ordinary way and completely ignore the procedure outlined in this section. However, it may pay him to read what follows, for a simpler method of computing the regression line is possible in a time series, owing to the fact that time values increase in equal increments.

15. Formula for the method of least squares. It is possible to describe a least squares trend line for a time series by the equation:

$$y = a + bd$$

where y is the value of the variable concerned, a and b are constants, and d is the *deviation of the required instant in time from the midpoint of the series*.

Note: Readers who skipped Chapter 13 may find Appendix 1:**5** useful here, if they read d for x.

This means that the problem becomes one of finding values for a and b. In this instance they are easily determined as follows:

a = arithmetic mean of the series
$b = \Sigma yd / \Sigma d^2$

Computing a is simple enough. The procedure for b is given below, although the experienced reader will be able to deduce the procedure from the formula.

16. Procedure for computing b. The procedure for computing b is as follows:

(*a*) Set down the annual figures in a vertical column.
(*b*) Find the midpoint of the series.
(*c*) Against each year, insert the deviation of the midpoint of that year from the midpoint of the series. (With an odd number of years in the series these deviations will be whole years, but with an even number they will involve half years.)
(*d*) Multiply the variable value for each year by its deviation and add the products to give Σyd.
(*e*) Square the deviations and add the products to give Σd^2.
(*f*) Apply formula given in **15**.

Example

Find the trend line of the data in Table 17A.

Year	y(sales)	d	yd	d^2
19–3	2,692	−1.5	−4,038	2.25
19–4	2,984	−0.5	−1,492	0.25
Midpoint				
19–5	3,348	+0.5	+1,674	0.25
19–6	3,712	+1.5	+5,568	2.25
	$\Sigma y = 12{,}736$		$\Sigma yd = +1{,}712$	$\Sigma d^2 = 5.00$

Now

$$a = \text{mean of the series} = \Sigma y/n = 12{,}736/4 = \underline{3{,}184}$$

and

$$b = \Sigma yd/\Sigma d^2 = +1{,}712/5 = \underline{342.4}$$
$$\therefore \text{Trend: } y = \underline{3{,}184 + 342.4d}$$

Note: The midpoint of the series is the *end* of 19–4. On the other hand each year's figures are, of course, *centred on the middle of the year*.

17. Use of the computed trend line. Having calculated the trend line, prediction is possible in one of two ways.

(a) The trend line can be graphed by taking any two years and:

(i) computing their deviations d_1 and d_2;
(ii) inserting these values of d_1 and d_2 in the trend formula and finding y_1 and y_2;
(iii) plotting d_1, y_1 and d_2, y_2 on the graph;
(iv) joining the two points to obtain a trend line. The trend line can then be extended and the estimated values in future years read off directly.

(b) Much more simply, take the year for which a prediction is required and:

(i) find its deviation d from the midpoint of the original series;
(ii) insert this value in the trend formula and compute y.

Example

Estimate the sales for 19–7 from the data in Table 17A.

19–7 has a deviation d of 2.5 years from the midpoint of the series in Table 17A.

$\therefore y = a + bd = 3,184 + 342.4 \times 2.5 = 4,040$ tonnes.

Note: Here the method of least squares gives to all intents and purposes the same answer as the method of semi-averages because the series of figures is a simple series and the added sophistication of the least squares is not really needed.

Random variations

Having isolated the seasonal variations and the trend in a time series we are left only with the random variations or, as they are sometimes called, the *residuals*. The method of finding these is quite a simple one – we merely compute the theoretical trend and superimpose the seasonal variations so as to obtain the theoretical figures for the different periods. And comparing these with the actual period figures gives us our random variations.

18. Subdividing the trend. Whilst the principle is very simple, its practical execution is complicated slightly by the need to subdivide the annual trend figure into figures for each seasonal period. If we call the trend figure for the *seasonal* period y', then the trend formula becomes:

$$y' = \frac{a}{n} + \frac{b}{n^2} \times d'$$

where n = number of seasons per annum, and d' = deviations measured in seasonal periods.

The trend figure for each period may now be computed by this formula.

Special care has to be taken in using the formula to count d' correctly. Remember that figures are centred at the midpoints of their periods, i.e. seasonal figures are centred at the midpoints of their respective seasons. Thus, if the deviation of quarter 3, 19–6 was required in our Table 17A example, it would be determined as follows:

Number of full quarters from midpoint of the series
(end of quarter 4, 19–4) to end of quarter 2, 19–6   = 6
Plus half a quarter for quarter 3, 19–6   = $\frac{1}{2}$

$$\therefore d' = 6\frac{1}{2}$$

19. Finding the random variation. Once we have the trend figure for any season we superimpose the seasonal variation by taking that trend figure and simply:

(*a*) *multiply* by the seasonal variation under the *multiplicative* method;

(*b*) *add* the seasonal variation under the *additive* method.

This gives the theoretical figure for the season which is then compared with the actual figure to find the random variation for that season. Repeating for all other seasons gives us all our random variations.

The procedure to find the random variations, then, is as follows:

(*a*) For each season:

(*i*) find d', the deviation (in seasons) of the season from the midpoint of the series;

(*ii*) find the trend for that season from the formula in **18**;

(*iii*) adjust for the seasonal variation (multiplying or adding according to the method selected);

(*iv*) find the difference between this and the actual figure for the season to obtain the random variation for that season.

(*b*) Repeat (*a*) for all seasons.

Note that if the random variations are given as a percentage of the theoretical figure the relative importance of these variations in the series can be seen.

Example

Find the random variations in our illustrative figures (use the multiplicative method).

The annual trend was found to be (**16**):

$y = 3,184 + 342.4\ d$ (where d = deviation in years)

Therefore the *quarterly* trend will be:

$$y' = \frac{3,184}{4} + \frac{342.4}{4^2} \times d' = 796 + 21.4d'$$

where d' = deviation in quarters.

The computation of the random variations is, therefore, as follows:

Quarter		d'	Trend (from formula above)	Seasonal variation (7)	Theoretical figure (Trend × Seasonal variation)	Actual figure	Random variation Actual − Forecast	%
19-3	1	$-7\frac{1}{2}$	636	104	661	672	11	+2
	2	$-6\frac{1}{2}$	657	95	624	636	12	+2
	3	$-5\frac{1}{2}$	678	100	678	680	2	0
	4	$-4\frac{1}{2}$	700	101	707	704	−3	0
19-4	1	$-3\frac{1}{2}$	721	104	750	744	−6	−1
	2	$-2\frac{1}{2}$	742	95	705	700	−5	−1
	3	$-1\frac{1}{2}$	764	100	764	756	−8	−1
	4	$-\frac{1}{2}$	785	101	793	784	−9	−1
Midpoint								
19-5	1	$+\frac{1}{2}$	807	104	839	828	−11	−1
	2	$+1\frac{1}{2}$	828	95	787	800	13	+2
	3	$+2\frac{1}{2}$	850	100	850	840	−10	−1
	4	$+3\frac{1}{2}$	871	101	880	880	0	0
19-6	1	$+4\frac{1}{2}$	892	104	928	936	8	+1
	2	$+5\frac{1}{2}$	914	95	868	860	−8	−1
	3	$+6\frac{1}{2}$	935	100	935	944	9	+1
	4	$+7\frac{1}{2}$	956	101	966	972	6	+1

These figures indicate that the random variations are very small and that the series is relatively stable.

Forecasting

Given that a series is acceptably stable it is possible to use our different measures to forecast future values. All that this involves is finding the future trend value for any season to be forecast, adjusting for the seasonal variation, and showing a potential margin of error based on the random variation. However, it should never be forgotten that any such forecast assumes that *no new factor will enter the series*.

20. Forecasting procedure. To prepare a forecast, then, the following procedure should be adopted:

(*a*) Analyse the past figures for the series concerned and identify the trend, seasonal variations and random variations.

(*b*) Compute the figures needed for the formula to subdivide the trend.

(*c*) For each season to be forecast:

(*i*) find the deviation of the season from the midpoint of the analysed series;

(*ii*) use the trend formula to find the trend for the season;

(*iii*) adjust the figure in (*ii*) for the seasonal variation;

(*iv*) state the figure in (*iii*) as the forecast, subject to an error based on the random variation.

Example

Forecast the 19–7 sales, quarter by quarter, for Fred's quarry.

The subdivided trend formula is that in **19**, i.e. $y' = 796 + 21.4d'$. Quarters 1, 2, 3 and 4 of 19–7 have deviations of $8\frac{1}{2}$, $9\frac{1}{2}$, $10\frac{1}{2}$ and $11\frac{1}{2}$ respectively from the end of 19–4, the midpoint of our analysed series. They also have seasonal variations of 104, 95, 100 and 101. If the random variations keep within the previous limits then our forecast should prove correct to within ±2%. Our forecast, then, is prepared as follows:

Fred's quarry sales – forecast for 19–7 (tonnes)

Qr	d'	Trend $796 + 21.4d'$	Seasonal variation	Point forecast	Random variation	Range forecast
1	$8\frac{1}{2}$	978	104	1,017	20	997 to 1037
2	$9\frac{1}{2}$	999	95	949	19	930 to 968
3	$10\frac{1}{2}$	1,021	100	1,021	20	1001 to 1041
4	$11\frac{1}{2}$	1,042	101	1,052	21	1031 to 1073

21. Interpolation and extrapolation. A common technique in statistics is to plot a series on a graph and then draw the line of best fit across the graph. The technique was used in connection with scattergraphs and regression lines and has now been applied to time series trends. The line of best fit is used to estimate values and it is in this context that we need to appreciate the importance, and significance, of interpolation and extrapolation. But first we must distinguish carefully between the two terms.

(*a*) *Interpolation* involves reading a value on that part of the line which lies *between* the two extreme points plotted, i.e. a value on the continuous line in Fig. 17.4.

(*b*) *Extrapolation* involves reading a value on the part of the line that lies *outside* the two extreme points plotted, i.e. a value on the dotted line in Fig. 17.4.

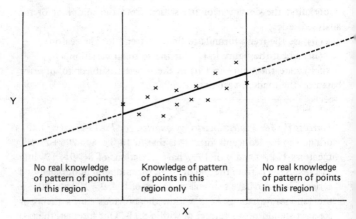

Figure 17.4 *Interpolation and extrapolation*

The distinction between the two is important, for although interpolation is permissible, it is considered dangerous to extrapolate. In the case of interpolation, the actual points on the graph give a sound indication of the possible error that could arise in reading a value from the line. But, where the line lies *outside* the plotted points, there is no guide at all to the degree of error. Although the plotted points may suggest the line has a steady slope, it may well be that some new, unsuspected factor comes into play at the higher or lower levels which, unknown to us, alters the slope in those regions.

For example, a scattergraph of heights and ages where ages are below 16 years would indicate continuous growth. If the height of an

80-year-old man was estimated by extrapolation from the line of best fit of such a scattergraph, the estimate would be something like 7.5 metres!

In this example, we know from experience that growth stops at about 16 years of age and that extrapolation would result in serious error. But in many cases we have no experience of the events lying beyond the two extremes (if we did we would have points to plot) and so therefore we do not know if extrapolating is allowable or not. It is true that the estimate may be reasonably accurate if it is read from a part of the line lying beyond, *but still close to*, one of the extreme points, but the further one moves away from the extreme point the more one must treat the estimate with caution.

Exponential smoothing

Exponential smoothing is a statistical averaging technique that enables a time series trend to be analysed. It is, in fact, no more than a special kind of moving average.

22. Defect of a normal moving average. As we have seen (5.**20**) a normal moving average is computed by adding the actual figures in a time series and then dividing by the number of periods. A little reflection will show that the earliest figure involved in computing such an average carries just as much weight as the latest. This is unfortunate, since the later figures will tend to reflect the more up-to-date average of the series and it will be better if they can be given more weight in the calculation than the earlier ones.

23. Exponential weighting. One way of achieving this desirable end would be simply to allocate straightforward progressively increasing weights to the figures in the series – for example, if we wanted a five-period moving average we could allocate weights of 1, 2, 3, 4 and 5 to the five consecutive figures involved in the computation. Note, incidentally, the last step would require us to divide the weighted total by the sum of the weights (*see* 10:**18**) – in this case by $1+2+3+4+5=15$.

Exponential weighting is, however, a neater and easier method of achieving such a progressive form of weighting. In this technique all one does each period is to weight suitably and then add together the current figure and the *previous period's moving average*. Then, providing that the sum of the weights is equal to 1 (and ensuring this is an essential aspect of the technique) this simple addition will give the new period's moving average without any further computation.

24. The smoothing constant. The key figure in this simply applied technique is called the *smoothing constant* (symbolised as α). The selection of a value for this constant gives the selector a measure of control over the degree of smoothing induced in the series. The value must, however, always lie between 0 and 1 (since the sum of the weights – of which this is one – must not exceed 1). The effect of different values is discussed below after the operation of the technique has been demonstrated.

25. Exponentially smoothing a series. A series is exponentially smoothed by applying the following formula:

New moving average = [current actual figure in series $\times \alpha$]
$$+ \text{[previous average} \times (1 - \alpha)]$$

Example

Exponentially smooth the series shown in column 2 of Fig. 17.5(a) using a smoothing constant α of 0.2.

Since the period 1 actual is 500, the period 'average' is obviously 500 too. Period 2, then, will be $(700 \times 0.2) + 500 \times (1 - 0.2) = 540$. The moving averages for the subsequent periods are similarly calculated and shown in the end column of Fig. 17.5(a).

26. Selecting the smoothing constant value. The selection of the value of α depends on how much weight it is desired to give later periods relative to earlier periods. As a quick calculation will show, giving α a value of 0 means the moving average will always be the first figure in the series (since all the later figures will be multiplied by 0 before being incorporated in the average), while a value of 1 means the average will always be the last figure (since all previous 'averages' will be multiplied by 0). This means that the higher α is, the greater the weightings of the later figures. The effect of using, for example, a smoothing constant of 0.8 in the series shown in Fig. 17.5(a) is demonstrated in Fig. 17.5(b).

(a) Smoothing a given series with a smoothing constant of 0.2.

Period	Actual figure	Exponentially smoothed moving average (to nearest whole number)
1	500	Actual figure $= 500$
2	700	$[700 \times 0.2] + [500 \times (1-0.2)] = 540$
3	600	$[600 \times 0.2] + [540 \times (1-0.2)] = 552$
4	550	$[550 \times 0.2] + [552 \times (1-0.2)] = 552$
5	900	$[900 \times 0.2] + [552 \times (1-0.2)] = 622$
6	1300	$[1300 \times 0.2] + [622 \times (1-0.2)] = 758$

(b) Smoothing previous series using a smoothing constant of 0.8.

Period	Average
1	Actual figure $= 500$
2	$[700 \times 0.8] + [500 \times (1-0.8)] = 660$
3	$[600 \times 0.8] + [660 \times (1-0.8)] = 612$
4	$[550 \times 0.8] + [612 \times (1-0.8)] = 562$
5	$[900 \times 0.8] + [562 \times (1-0.8)] = 832$
6	$[1300 \times 0.8] + [832 \times (1-0.8)] = 1206$

(c) Graph of actual values in the series and the two exponentially weighted moving averages.

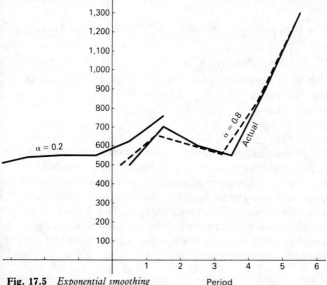

Fig. 17.5 *Exponential smoothing*

Although a high smoothing constant keeps the average more closely aligned to the actual figures it suffers the disadvantage that it does little to smooth the series and is unduly influenced by random variations. Consequently in practice α is generally chosen so that it lies between 0.1 and 0.3 – the final choice often being made by testing different values in conjunction with past data and observing which 'works' best.

27. Sum of the weights in exponential smoothing. It should be appreciated that by weighting the current actual figure with α and the previous average with $(1 - \alpha)$, the sum of the weights is $\alpha + (1 - \alpha) = 1$. So, since the sum of the weights is 1 and dividing any figure by 1 results in the figure being unchanged, the moving average is simply the sum of the two weighted figures.

28. Plotting an exponentially smoothed average. Up to now we have plotted moving averages at the midpoint of the period to which they relate. Where, however, can one plot an exponentially smoothed moving average?

The answer to this question is that the average must be plotted with a $\dfrac{1-\alpha}{\alpha}$ period *lag* (i.e. prior to the current period). Thus, when α is 0.2, the average must be plotted with a $\dfrac{1-0.2}{0.2} = 4$ period lag. Similarly, an α of 0.8 requires a $\dfrac{1-0.8}{0.8} = \dfrac{1}{4}$ period lag. Fig. 17.5(c) shows the plotting of our previously calculated two sets of averages together with the original series.

29. Forecasting from an exponentially smoothed average. If it is desired to forecast figures from an exponentially smoothed average then it is first necessary to determine a period trend value. Once this is done by whatever technique the forecaster may elect to use, the forecast is made by first extrapolating the last average up to the current period (i.e. adding *trend value × (1 − α)/α* to the last average) and then extrapolating on to the required future period (i.e. further adding *trend value × number of periods that the future period lies ahead*).

Example

In our illustration the rise in the trend found by the method of semi-averages is $(500+700+600)/3$ to $(550+900+1,300)/3$ over the time span from the midpoint of periods 1, 2 and 3 and the periods 4, 5 and 6, i.e. 600 to 917 from middle of period 2 to middle of period $5 = 317/3 = 106$ (rounded) per period.

Now the last weighted average figure was 758 which was plotted $(1-0.2)/0.2 = 4$ periods behind the current period. So extrapolating this figure up to the current period gives $758 + (106 \times 4) = 1182$.

A forecast for n periods ahead, therefore, would be given by the formula $1182 + (106 \times n)$.

Progress test 17

(Answers in Appendix 4)

1. The data (1) on p. 228 relates to the populations, in thousands, of two towns A and B, between 1978 and 1986.

(a) Compute the population trend line for each town.
(b) Estimate when the populations of the two towns will be equal, and state the estimated size of the populations at that time.

2. (a) From the figures (2) on p. 228 compute (i) the seasonal variations, and (ii) the trend.
(b) Estimate (i) the total demand for 19-8 and (ii) the demand (using the multiplicative method) in the last quarter of 19-8.

Year	A	B
1978	25	200
1979	31	196
1980	36	194
1981	44	190
1982	48	189
1983	53	185
1984	60	184
1985	62	181
1986	67	180

Data 1

Demand figures (tonnes) 19–4 to 19–6

Year	Quarter 1	Quarter 2	Quarter 3	Quarter 4
19–4	218	325	273	248
19–5	444	585	445	385
19–6	660	852	623	525

Data 2

3. (*For readers who have studied regression lines*) (*see* Chapter 13).

Prove that a trend line found by the method of least squares is the regression line of y on x, where x is time. (*Hint:* If deviations are measured from the midpoint of a series and such deviations are in equal steps, note that the sum of these deviations must be zero.)

4. Exponentially smooth (to two decimal places) the following series of period figures using a 0.25 smoothing constant: 8, 12, 15, 14, 17, 20, 25, 23, 28, 30, 33.

Assignment

1. The following figures relate to units of service demanded of a service enterprise working 24 hours a day.

Day	Morning	Afternoon	Night
1	820	310	600
2	800	330	600
3	860	340	700
4	900	380	680

(a) Find (i) the seasonal variation, (ii) the trend, and (iii) the random variation.

(b) Graph the actual figures and superimpose the trend line.

2. (a) Find the seasonal variation in the series in Assignment 2 in Chapter 5.

(b) Forecast the figure for the fourth quarter of the year following the end of this series.

Part six
Sampling theory and inference

18
Probability: 1

Probability usually gives students more trouble than any other topic. It is certainly a field rich in examples of erroneous deductions, even among mathematicians. There seems to be no short cut to gaining an understanding of the subject and the reader should, therefore, work carefully through this chapter and the progress tests. If there is one hint that may prove useful it is that the first crucial step in solving a probability problem is to be very clear just what the problem is, for if there is ambiguity in the problem itself there is a serious danger that the answer will be flawed. For instance, if we have two presses in a particular department and ask, 'What is the probability one of the two presses will break down next month?', then is the answer the probability of just a single machine breaking down or will it include the probability that both will break down? Perhaps in no other topic is an accurate statement of the problem such a significant step towards the final solution.

Basic probability theory

First we look at the most basic principles of probability theory, upon which all the further theory is built. It should perhaps be made clear that the understanding of all probability discussion is very dependent upon grasping each and every prior step and so the reader is advised to read this section particularly carefully.

1. Probability. Strangely, there seems to be no way to define probability that does not, to start with, rely on our understanding of what the word means. Since uncertainty is fundamental to the universe we live in we seem to understand intuitively the concept of probability. Unfortunately intuition can sometimes be deceptive and lead us astray. It is likely that it is this intuitive element that gives us so much difficulty in understanding probability theory and so the reader is advised to be sure that he can show that his answer to a probability question is logically sound.

2. Measuring probability – method of relative frequency.
Even defining a measure of probability is not free from problems but, for all practical purposes, we can measure probability by expressing as a fraction how often an event occurs in relation to how often it could occur. Thus, if we toss a coin we would expect a 'head' to occur about half as often as it *could* occur, and if we threw a die we would expect to throw a 6 on about one-sixth of the throws. We have, therefore, probabilities of half (½) of tossing a 'head' and 1/6 of throwing a 6. Putting it another way, we can say the probability of an event is the long-term average of the number of occurrences of the event relative to the number of attempts – e.g. if we toss a coin often enough, the average number of 'heads' will be one-half the number of tosses. This method of measuring probability is sometimes referred to as the *method of relative frequency*. The full range of probabilities is diagrammatically shown in Fig. 18.1.

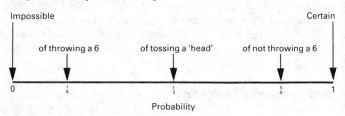

Note: No event can have a probability of less than 0 (since an event cannot occur less often than never) or more than 1 (since an event cannot occur more often than every time).

Figure 18.1 *Measuring probability*

3. Events.
If ambiguity is the bane of probability analysis then careful definitions are the keys to removing ambiguity. In this section the basic definitions involving *events* are given and the student should learn these thoroughly.

(a) *Event.* An event is simply an occurrence in a defined context. So to toss a 'head' is an event (where the context is 'one coin tossed'), as is tossing three 'heads' with three coins (where the context is 'three coins tossed'). Another term for event is *outcome.*

(b) *Elementary event.* An elementary event is an event which cannot be broken down into sub-events. Thus, tossing a 'head' is an elementary event since this cannot be broken down into any sub-events. Conversely, tossing three 'heads' with three coins is *not* an elementary event as this can be broken down into tossing a 'head' on three separate occasions, i.e. into three sub-events.

(c) *Compound event.* A compound event is an event which comprises two or more elementary events – such as tossing three 'heads'.

(d) *Sequential event.* A sequential event is an event that comprises a *sequence of elementary events.* So, as well as a compound event, tossing three 'heads' is also a sequential event. Clearly, there is a similarity between a compound event and a sequential event. The difference, however, lies in the fact that in a sequential event the *sequence* of the elementary events is crucial. Thus, tossing three coins so that the first falls a 'head', the second a 'head' and the third a 'tail' (HHT) is a sequential event as is the case where the first toss falls a 'tail', the second a 'head' and the third a 'head' (THH). We also have the separate sequential event HTH. But the event two 'heads' and a 'tail' is a compound event. As can be seen, then, a single compound event may comprise a number of different sequential events.

(e) *Mutually exclusive events.* Mutually exclusive events are events that *cannot* occur together. Thus, obtaining a 'head' and a 'tail' with a single toss of a coin are mutually exclusive events since one cannot have both at the same time. Conversely, tossing a 'head' and throwing a 6 with a die are not mutually exclusive events since they can occur together. Note that in the case where we toss three coins the sequential events (HHT, THH, HTH) are all mutually exclusive events since three tosses of a coin to give two 'heads' and a 'tail' can be the result of only one of these three events. The importance of this point is shown in **9**.

(f) *Equiprobable events.* Equiprobable events are events which are all equally probable. Thus, it is equally probable that a coin will come down 'heads' as it is that it will come down 'tails'. And it is equally probable that the ace of clubs will be drawn at random out of a pack of cards as the king of diamonds.

(g) *Complementary events.* Complementary events are events where one or other is certain to occur, but not both together. Thus, if a single coin is tossed the event *'heads'* and the event *'tails'* are complementary.

4. Trial. Another important term is trial. A *trial* is a single attempt to obtain a defined event. Thus, if an attempt is made to obtain a 'head' in a single toss of a coin, then a single toss of a coin is a trial. But if an attempt is made to obtain a head on three successive tosses, then the trial comprises three successive tosses.

5. Empirical and a priori probabilities. There are two ways that the probability of an *elementary* event can be found. These are the empirical and the a priori. To illustrate these ways, assume that there is a board with parallel lines 5cm apart and that we want to know the probability of a disc 2cm in diameter, thrown at random on the board, touching one of the lines (*see* Fig. 18.2(*a*)).

(a) *Empirical probability.* An empirical probability is one that is established by *observing* the relative frequency of trial successes to total trials. In the case of our illustrative problem we will find the required probability by throwing the disc on the board a large number of times and observing each time it cuts a line. After, say, 1,000 throws we will estimate the probability that the disc will cut a line by dividing the number of times this happens by 1,000, the total number of throws. In real life the empirical method is often the only way we can estimate probabilities. So insurance companies find the probability, say, of a house being burned down by observing the number of times a house they have insured has burned down relative to the total number of houses they have insured. By classifying such houses they can determine how much greater the probability is of a house with, say, a thatched roof burning down than a house with a slate roof.

(b) *A priori probability.* An a priori probability is one that is established by an a priori belief (i.e. intuitive and generally acceptable belief) as to what, in the absence of any trials whatsoever, the relative number of successes to the total number of trials must be. In our illustrative problem we would doubtless intuitively assert that the disc

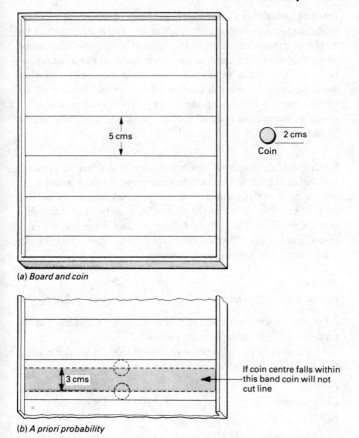

(a) Board and coin

If coin centre falls within
this band coin will not
cut line

(b) A priori probability

Figure 18.2 *Empirical and a priori probabilities*

would be as likely to fall in one position as another (i.e. the events are equiprobable). Since the disc will not cut a line if its centre falls beyond 1cm of any line, and since there is 5cm between the lines, then there is a band 3cm wide within which the centre can fall without cutting a line. And so the probability of the disc not cutting a line is 3/5 (*see* Fig. 18.2(*b*)). Note that a priori probabilities often assume that events are equiprobable.

Generally on those occasions where both methods can be used the resulting probabilities are in agreement but there can be occasions when they are not (e.g. a magnet under our board may result in the empirical probability being very different from our 3/5 a priori probablity). In such a case it is not easy to decide if our intuition is at fault or if the results we observe when using the empirical method were, by chance, one-sided, or even if we have fully understood the circumstances surrounding the events. The problem, therefore, should continue to be analysed until the two kinds of probabilities can be reconciled.

6. Empirical method computations. Normally, finding probabilities using the empirical method simply involves counting the total number of trials and the number of occurrences of the defined event. Sometimes, however, the probabilities can be computed from a table of numbers. Look at the following table, for instance, which shows out of an initial batch of 1,000 units the number of rejects at the end of each of four consecutive processes:

Process	1	2	3	4
Rejects at end of process	150	120	210	50

For each process, what are the probabilities that a unit entering the process will move on to the next process?

To compute these probabilities we must divide the number starting the process by the number successfully completed, i.e.

Process	1	2	3	4
Units starting	1000	850	730	520
Rejects	150	120	210	50
Units successfully completed	850	730	520	470
Probability of successful completion	$\frac{850}{1000}$	$\frac{730}{850}$	$\frac{520}{730}$	$\frac{470}{520}$

Other probabilities can similarly be found, e.g. the probability of a unit at the start successfully completing all processes = 470/1000, and the probability that a unit that successfully completes process 1 will be rejected at process 3 = 210/850.

7. Probability symbols. To give clarity and brevity to probability expressions mathematical type symbols are often employed. Those immediately relevant to this chapter are as follows:

(a) *P(. . .)*. This is the symbol for the probability of whatever event is stated between the brackets. So, when tossing a coin, P(Head)=½, and P(Tail)=½ as well.

(b) *P(A∩B)* The symbol ∩ is read as AND, so P(A∩B) is the probability of both events A AND B. So if we roll a die and toss a coin the probability of throwing a 3 AND tossing a 'head' is shown as P(3∩Head).

(c) *P(A∪B)*. The symbol ∪ is read as OR, so P(A∪B) is the probability of A OR B. So the probability of throwing a 3 with a die OR tossing a 'head' is P(3∪Head).

(d) *P(Ā)*. The bar over the symbol for an event indicates the probability that it will *not* occur. Where there are two complementary events P(Ā) is the probability that A's complement will occur. Note that sometimes a dash is written instead of a bar, i.e. $A' = \bar{A}$.

8. Principles of basic probability computations. Examination questions are almost always concerned with compound events. And the secret of tackling such questions is to *analyse the compound event into a set of sequential events*.

Taking a simple illustration, say we wish to know the probability of obtaining two 'heads' and a 'tail' from a toss of three coins. To solve this we analyse the compound event into the sequential events HHT, THH and HTH. Once we have the sequential events, finding the probability of a compound event becomes the relatively straightforward application in turn of what are called the multiplication rule and the addition rule.

(a) *Multiplication rule*. The multiplication rule says that the probability of a sequential event is the *product* of all its elementary events – i.e. P(A∩B∩C∩D . . .) is P(A)×P(B)×P(C)×P(D) . . . So the probability of HHT (where P(Head)=½ and P(Tail)=½) is ½×½×½=⅛. Also the probability of THH=½×½×½=⅛ and the probability of HTH=½×½×½=⅛.

(b) *Addition rule*. The addition rule says that the probability of one of a number of *mutually exclusive* events occurring is the *sum* of the probabilities of the events, i.e.

$$P(Z \cup Y \cup X \cup W \ldots) = P(Z) + P(Y) + P(X) + P(W) \ldots \text{ etc.}$$

In our instance, therefore, the probability of the compound event 'two heads and a tail' is the sum of the probabilities of the mutually exclusive sequential events HHT, THH and HTH – i.e. $\frac{1}{8} + \frac{1}{8} + \frac{1}{8} = \frac{3}{8}$.

9. Illustrative basic probability computations. Because the theory of the previous paragraph is so important it is worthwhile looking in this paragraph at one or two additional computations.

(a) A card is taken from two separate packs. What is the probability that at least one card is a spade?

This is a compound event which can be analysed into three sequential events with their individual probabilities (note: P(spade) = $\frac{1}{4}$, P(not spade) = $\frac{3}{4}$):

Spade ∩ spade:	Probability = $\frac{1}{4} \times \frac{1}{4} = 1/16$
Spade ∩ not spade:	Probability = $\frac{1}{4} \times \frac{3}{4} = 3/16$
Not spade ∩ spade:	Probability = $\frac{3}{4} \times \frac{1}{4} = 3/16$

Since these are mutually exclusive events the probability of one or the other of them occurring is the sum of their individual probabilities. So the probability of at least one spade = 1/16 + 3/16 + 3/16 = 7/16. And if we wanted the probability of exactly one spade then we would select the sequential events that gave us just one spade and add their probabilities – i.e. 3/16 + 3/16 (the bottom two sequential events) = $\frac{3}{8}$.

Note, incidentally, that when forming a sequential event it is important to include the probabilities of the elementary complementary events as well as the probabilities of the events in which we are interested. Note, too, the question specifies a card from two separate packs. If the same pack were used and the first card selected were the spade, then the probability of the second card being a spade would not be $\frac{1}{4}$ but 12/51 (since there would be only 12 spades in a pack of 51 cards – i.e. a relative frequency of 12/51).

(b) What is the probability of scoring ten or more with a throw of two dice?

This compound event can be broken up into the following mutually exclusive sequential events:

First Die	Second Die	Score	Probability
6	6	12	$1/6 \times 1/6 = 1/36$
6	5	11	$1/6 \times 1/6 = 1/36$
6	4	10	$1/6 \times 1/6 = 1/36$
5	6	11	$1/6 \times 1/6 = 1/36$
5	5	10	$1/6 \times 1/6 = 1/36$
4	6	10	$1/6 \times 1/6 = 1/36$

Probability of compound event '10 or more'	$= 6/36$
	$= 1/6$

Further examples will also be found in the progress test.

10. Cross-check to sequential event probabilities. It should be appreciated that if we list *all* the possible sequential events in a situation then, since it is certain one of them must occur, the sum of their probabilities must come to 1. This, then, enables us to make a cross-check, for if we add the probabilities of all the possible sequential events and obtain 1 as a total then we know that we have:

(*a*) Ensured that all our sequential events are genuinely mutually exclusive.

(*b*) Computed correctly the probabilities of the individual sequential events.

For example, in our card illustration only one possible sequential event remains to be listed – the 'not spade ∩ not spade' event. The probability of this event is $3/4 \times 3/4 = 9/16$, and adding this to our previous probability total of 7/16 gives us $9/16 \times 7/16 = 1$. This proves the accuracy of our earlier work.

11. 1 – complementary event. Finally it should be appreciated that since the sum of the probabilities of all the sequential events is 1, then a very useful short-cut is available to us where the event we want comprises the large majority of the sequential events – all we need to do is to take the probability of the complementary or the alternative events from 1. So:

Probability of at least 1 spade $= 1 -$ probability of no spades at all
$$= 1 - P(\text{Not spade} \cap \text{Not spade})$$
$$= 1 - (\tfrac{3}{4} \times \tfrac{3}{4}) = 1 - 9/16 = 7/16$$

Progress test 18

(*Answers in Appendix 4*)

1. Explain carefully each of the following terms:

(*a*) method of relative frequency (**2**);
(*b*) elementary event (**3**);
(*c*) compound event (**3**);
(*d*) sequential event (**3**);
(*e*) mutually exclusive events (**3**);
(*f*) equiprobable events (**3**);
(*g*) complementary events (**3**);
(*h*) trial (**4**);
(*i*) empirical probability (**5**);
(*j*) a priori probability (**5**).

2. In calculating the premiums payable for life assurance policies insurance companies use information derived from mortality tables. Extracts taken from a mortality table are given below:

	Males per 1,000 births
Age in years	*Number living*
0	1,000
25	958
45	905
65	680
75	413

Find the probability that a male aged 25 will:

(*a*) attain the age of 45 years;
(*b*) attain the age of 45 years but not 75;
(*c*) not attain the age of 65 years.

(*CIMA*)

3. A proposal, to put more emphasis on improved superannuation rather than wage increases, was put by the union executive to the members. The following results were received:

	Opinion			
	In favour	*Opposed*	*Undecided*	*Total*
Members:				
skilled	800	200	300	1,300
unskilled	100	600	200	900
	900	800	500	2,200

Calculate the probability that a member selected at random will be:

(*a*) skilled and in favour of the proposal;
(*b*) undecided;
(*c*) either unskilled or opposed to the proposal.

(CIMA)

4. What is the probability of:

(*a*) (*i*) tossing two 'heads';
    (*ii*) throwing a double six;
(*b*) Throwing a die and scoring a 5 or 6.
(*c*) Drawing an ace from a pack of cards.
(*d*) Tossing a 'head' and a 'tail'.
(*e*) Throwing two dice and scoring a total of 11.
(*f*) Throwing two dice and scoring 7.

5. The probability that a man now aged 55 years will be alive in 1993 is 5/8, while the probability that his wife now aged 53 will be alive in 1993 is 5/6. Determine the probability that in 1993:

(*a*) both will be alive;
(*b*) at least one of them will be alive;
(*c*) only the wife will be alive.

(CIMA)

6. In the past, two building contractors, A and B, have competed for 20 building contracts of which 10 were awarded to A and 6 were awarded to B. The remaining 4 contracts were not awarded to either A or B. Three contracts for buildings of the kind in which they both specialise have been offered for tender.

Assuming that the market has not changed, find the probability that:

(a) A will obtain all three contracts;

(b) B will obtain at least one contract;

(c) Two contracts will not be awarded to either A or B;

(d) A will be awarded the first contract, B the second, and A will be awarded the third contract.

(CIMA)

7. The probability of a company obtaining contract A is $0 \cdot 3$, contract B $0 \cdot 4$ and contract C $0 \cdot 6$. If the company obtains none or only one of these contracts it will bid for D with a $0 \cdot 8$ probability of obtaining it. What is the probability that the company will obtain:

(a) just A and B;

(b) at least A and B;

(c) C and D;

(d) just C;

(e) just D.

8. A company has a security system comprising four electronic devices (A, B, C and D) which operate independently. Each device has a probability of failure of $0 \cdot 1$. The four electronic devices are arranged so that the whole system operates properly if at least one of A or B functions, *and* at least one of C or D functions.

What is the probability that the whole system operates properly?

(CIMA)

9. A market survey worker is required to interview a quota of two retailers each day. Unfortunately retailers are not always prepared to be interviewed – in fact, there is a $0 \cdot 4$ probability that a given retailer will refuse. If the worker can approach a maximum of four retailers in a day, what is the probability that:

(a) he will have completed his quota after visiting:

(i) 2 retailers?

(ii) 3 retailers?

(iii) 4 retailers?

(b) he will fail to achieve his quota?

10. The delivery of an item to a company from a supplier can take either 1, 2 or 3 weeks from the order date. On average, 20 per cent of deliveries take 1 week from ordering, 50 per cent take 2 weeks and 30 per cent take 3 weeks. The company uses either 1 or 2 items each

week, with $0 \cdot 6$ and $0 \cdot 4$ probabilities, respectively. The number used in one week is independent of how many have been used in previous weeks.

(a) (i) What is the probability that a delivery from the supplier takes 2 weeks or longer?

(ii) Given that the company has not received delivery in the first week after ordering, calculate the probability that it will receive delivery during the second week, and interpret your answer.

(iii) Calculate the probability that the company uses

        (1) 3 items in two weeks;

        (2) 4 items in two weeks.

(b) If the company has 2 items in stock when it places an order for delivery, calculate the probability that it will have insufficient stock to meet production before the delivery arrives.

(ACCA)

Assignment

Construct a board similar to that in Fig. 18.2(a) and by throwing a coin on to it at random a large number of times, compute the diameter of the coin.

19
Probability: 2

Now that we have the fundamentals of probability theory we can look at different aspects of the subject. In this chapter a number of general topics are first discussed before turning to the topic of conditional probability.

General probability topics

The previous chapter was concerned in presenting fundamental concepts in probability theory. Here we look at topics associated with, but less fundamental to, that theory.

1. The law of averages. There is one misconception that should be cleared up right at the beginning of our study of probability and that is the view that what is often called the 'law of averages' will ensure that, if a run of trials favours one type of event more than it theoretically should, then soon the 'neglected' events will score relative to that previously favoured.

The law of averages is really the law of large numbers and what it says is that in the long term the proportion of times a result actually occurs will approach closer and closer to its theoretical proportion. Take, for instance, tossing a coin. The law says that in the long run the proportion of 'heads' will be half the total tosses. But if we have a long run of 'tails' – say 80 'tails' in 100 tosses, a 0.8 proportion against the theoretical 0.5 proportion – we do *not* need a subsequent predominance of 'heads' in order to 'comply' with the law. Say, for instance, in the next 900 tosses we have 450 'tails' and 450 'heads'. In this case the total 'tails' will be $80 + 450 = 530$ in 1000 trials, i.e. a proportion of 0.53. So even though the 'heads' and 'tails' are equal in the last 900 tosses, nevertheless the proportion of 'tails' drops from 0.8 to 0.53. Indeed, 'tails' can still predominate (500 to 400 'heads' in the last 900 tosses) and yet the proportion will still fall (500 + 80 = 580 in a 1000 which is a proportion of 0.58). All this means is that while the law of large numbers is perfectly valid, the interpretation that many gamblers put upon it is wholly erroneous.

2. Sample space. We have seen that the procedure for finding the probability of a compound event is to analyse it into a set of sequential events, each comprising elementary events (18:**8**). We have also seen that if we list all the possible sequential events that arise in the situation then the sum of the events probabilities will be 1 (18:**10**). Such a complete list is obviously useful and warrants its own term. This term is *sample space* and, formally, we have the following definitions:

    (*a*) *Sample point.* A sequential event.
    (*b*) *Sample space.* The set of all the sample points.
    (*c*) *Compound event.* A subset of the sample space.

3. Success and failure. In probability theory, if the event with which we are concerned occurs, we say we have a *success*. If, then, we are concerned with the die falling a 6, each time it falls a 6 we have a success. (Paradoxically, if we are concerned with the number of people who fail an examination, every failure is a success!) The probability of a success is symbolised by p and so we can say that in the case of a die falling a 6 then $p = 1/6$th.

If the event we are concerned with does not occur we have, of course, a *failure*. The probability of a failure is symbolised as q. As a moment's study of the probability scale in Fig. 18.1 will quickly show $q = 1 - p$, i.e. the probability of a failure is always '1 minus the probability of a success'. Note, that q can also be defined as the *probability of the opposite event* and that $p = 1 - q$.

4. Probability of at least one success. The fact that $p = 1 - q$ can often be used to save computational work. For instance, assume that we want the probability of at least one 'head' when five coins are tossed together. We could solve this by finding the probability of one 'head', two 'heads', three 'heads', four 'heads', and five 'heads' and then adding these probabilities (since they are all mutually exclusive events). However, note that if at least one 'head' is a success, zero 'heads' is the only failure. And the probability of zero 'heads' is $P(T \cap T \cap T \cap T \cap T) = \frac{1}{2}^5 = 1/32$. Since $p = 1 - q$ then the probability of a success (i.e. at least one 'head') must be $1 - 1/32 = 31/32$.

So generally we have:

P(compound event occurring) = 1 − P(compound event not occurring)

5. Probability trees. When there are only two or three elementary events in the sequential event a complete listing of all the sample points in the entire sample space is a simple enough task. Where, however, the number of elementary events increases, so does the difficulty of ensuring each and every sequential event is correct. The probability tree is a technique designed to overcome this difficulty, and is best explained by means of an illustration.

Assume that the die has four red and two yellow faces. We throw the die four times. What is the probability we will obtain a total of two red and two yellow faces?

To construct the appropriate probability tree for this problem we start at START on the left-hand side of the tree (*see* Fig. 19.1) and then draw 'branches' out from it to circles representing each possible result we can obtain from the first throw of the die (two in this illustration – a red face and a yellow face). The probability of each result is then written alongside the branch leading to that result. Then, considering *each new circle in turn* we again draw branches, with the appropriate probability shown, to each possible alternative at the next throw of our die. This procedure continues until we reach the end of the experiment – i.e. after a fourth throw of our die.

The tree is now constructed and all that remains is to list the sequential events and compute their probabilities. The sequential events are listed by tracing the order from START to the end of each branch and the sequential event probabilities are found by multiplying together the probabilities marked against each branch traversed. Finding the probability of obtaining the compound event two reds and two yellow faces, with our four throws, then, simply involves identifying the sequential events that form that compound event subset – i.e. we count the successes – and adding their

probabilities. As can be seen from Fig. 19.1, this gives us
$(4+4+4+4+4+4) \div 81 = 8/27$.

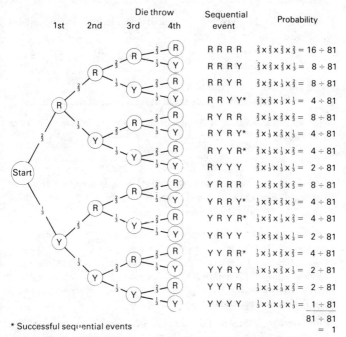

Figure 19.1 *Probability tree*. 4 throws of a die with 4 red faces and 2 yellow faces.

6. Expectation. So far most of our events have had no 'value' – a red face falling was just a red face and we made no attempt to value its occurrence. However, if all the events of a probability situation have values then it becomes possible to compute what the average value would be if the situation repeated itself frequently. This average is termed the *expectation* (or, more fully, *mathematical expectation*). This concept sounds perhaps more complex than it is for the reader will doubtless appreciate that if he were paid £1 every time he tossed a 'head' with a coin then he would win on average 50p a toss. So his expectation would be 50p a toss (and £5 if he were allowed a sequence of ten tosses of the coin).

Computing the expectation in any probability situation is hardly any more difficult than grasping the concept – all that needs to be done is to:

(*a*) list all the possible sequential events (the sample space) appertaining to the situation;

(*b*) value each sequential event (sample point);

(*c*) apply the formula:

Sequential event expectation
= probability of sequential event × value of sequential event

(*d*) add the sequential event expectations to obtain the total expectation.

Example

A die having four red and two yellow faces is thrown twice. For each red face showing the thrower is paid 9p and for each yellow face 27p. What is the thrower's expectation?

Sequential event	Prob. of event	Value of event(p)	Expectation(p)
red, red	$2/3 \times 2/3 = 4/9$	$9 + 9 = 18$	$4/9 \times 18 = 8$
red, yellow	$2/3 \times 1/3 = 2/9$	$9 + 27 = 36$	$2/9 \times 36 = 8$
yellow, red	$1/3 \times 2/3 = 2/9$	$27 + 9 = 36$	$2/9 \times 36 = 8$
yellow, yellow	$1/3 \times 1/3 = 1/9$	$27 + 27 = 54$	$1/9 \times 54 = 6$

∴ Thrower's expectation = 30p $\qquad \overline{30}$

On average, then, the thrower will win 30p. (Note, incidentally, that an expectation need not be, and in fact rarely is, an *actual* value.)

Conditional probability

So far the probability of the elementary events we have looked at has not changed as the analysis has proceeded. So the probability of tossing a 'head' has been 0.5 regardless of everything else that may have happened. This, however, is not always the case. For instance, the probability of someone breaking his leg will depend on what he is doing – skiing or sleeping, climbing or swimming. When the probability of an event depends upon the condition surrounding it we have *conditional probability* and in this section we look at this aspect of probability theory.

7. Statistically independent events. *Statistically independent events* are events which are in no way affected by the occurrence of each other. Thus, if a die were thrown and a coin tossed the probability of a 'head' being tossed is unaffected by the outcome of the die throw, and vice versa. The results of the toss and the throw, therefore, are statistically independent events. Many events, however, are dependent on other events. So, as we saw above, the probability of a broken leg depends very much on what is being done at the time. Such events are not statistically independent events.

8. Prior and subsequent events. When two events are not statistically independent then usually the probability of one event is governed by the other event which is itself independent – i.e. there is a parallel between events and dependent and independent variables. Thus, the probability of getting wet in a rain shower depends on whether it rains or not – and whether it rains or not does not depend on how wet you are. To distinguish one event from the other we use the terms *prior* and *subsequent* events:

  (*a*) *Prior event.* The event which is independent of the other event.
  (*b*) *Subsequent event.* The event which is dependent on the prior event.

Although a prior event is usually recorded before the subsequent event (rain is observed before you are seen to be wet) this is by no means invariable. If you walked into a deep underground room soaked to the skin the first event observed (recorded) by someone inside would be your wetness. Only after that could it be deduced that it had been raining – the prior event.

9. Conditional probability. A *conditional probability* is an event probability which is conditional on another event. Assume there is an urn containing four red and six yellow marbles. (By tradition urns are regarded as indispensable to the study of probability.) Two draws are made (without replacing the drawn marbles). What, after the first draw, is the probability that the second marble drawn is red?

Clearly, this probability depends very much on, i.e. is conditional upon, the colour of the first marble. If this were red, then the probability of the next marble drawn being red would be 3/9 (3 red marbles out of 9), but if it were yellow the probability would be 4/9.

These two probabilities are therefore called conditional probabilities since they are conditional on the result of the first draw.

10. P(A|B). In conditional probability theory we use the symbol P($\ldots$|$\ldots$) to express a conditional probability, the subsequent event being written in front of the vertical line inside the brackets and the prior event after. The vertical line stands for 'given that'. So, the probability of being soaked to the skin should it rain is symbolised as P(soaked|rain) – i.e. probability of being soaked given that there is rain. And the two results in the previous paragraph could be written P(red|red) = 3/9 – i.e. the probability of drawing a red given a red already drawn – and P(red|yellow) = 4/9 – i.e. the probability of drawing a red given a yellow already drawn.

To summarise:

P(subsequent event|prior event) = probability of subsequent event given prior event occurred.

11. Conditional probability analyses. When making conditional probability analyses the multiplication and addition rules we looked at earlier (*see* 18:8) remain exactly the same. Only the elementary subsequent event probabilities change according to what prior events occur.

Example

Given an urn containing four red and six yellow marbles, find the probability of obtaining two red and two yellow marbles if four marbles are drawn one after the other. (*See* p. 253)

12. P(subsequent event|any prior event). Very often we are only interested in the probability of a subsequent event and are indifferent to whatever prior event may have preceded it. In such a case all we do is:

(*a*) list all the sequential events which *end* with the required subsequent event;
(*b*) find the probability of each sequential event;
(*c*) add the probabilities.

Example

My son is to meet me by a given time. The probability of his being

Compound event	Sequential event	Probability	
	RRYY	$\frac{4}{10} \times \frac{3}{9} \times \frac{6}{8} \times \frac{5}{7} =$	$360 \div 5{,}040$
	RYRY	$\frac{4}{10} \times \frac{6}{9} \times \frac{3}{8} \times \frac{5}{7} =$	$360 \div 5{,}040$
2R & 2Y	RYYR	$\frac{4}{10} \times \frac{6}{9} \times \frac{5}{8} \times \frac{3}{7} =$	$360 \div 5{,}040$
	YRRY	$\frac{6}{10} \times \frac{4}{9} \times \frac{3}{8} \times \frac{5}{7} =$	$360 \div 5{,}040$
	YRYR	$\frac{6}{10} \times \frac{4}{9} \times \frac{5}{8} \times \frac{3}{7} =$	$360 \div 5{,}040$
	YYRR	$\frac{6}{10} \times \frac{5}{9} \times \frac{4}{8} \times \frac{3}{7} =$	$360 \div 5{,}040$

$\therefore$ Probability 2 reds and 2 yellows = $\underline{\underline{2{,}160 \div 5{,}040}}$

punctual if he travels by bus is 0.2, if he travels by train 0.5 and if he is given a lift in the neighbour's car 0.9. The probabilities of him travelling by bus, train and car are 0.3, 0.4 and 0.3 respectively. What is the probability that he will be on time?

Solution

Here I am indifferent as to just how my son travels – I am only interested in whether he will arrive on time or not. So the analysis will be (distinguishing between prior and subsequent events and applying the conventional symbolism):

Sequential event

Prior event	Subsequent event	P(prior)	P(subs|prior)	P(subsequent)
Bus	arrive on time	0.3	0.2	0.06
Train	arrive on time	0.4	0.5	0.20
Car	arrive on time	0.3	0.9	$\underline{0.27}$
		P(arrive on time|any prior event) =		$\underline{0.53}$

So the probability he will arrive on time is just over ½.

13. The probability square. A useful, if unorthodox, visual device for grasping conditional probabilities is what can be called the *probability square*. This is simply a square with sides one unit long (e.g. one inch). The bottom side is divided into sections equal to the probabilities of all the possible *prior* events, and verticals drawn to divide the square into a series of adjoining rectangles. Across each rectangle a horizontal line is drawn at a height equal to the conditional probability of the *subsequent* event given the prior event associated with the rectangle. The total area, then, lying below all the horizontal lines is the total probability of the subsequent event – i.e. *P(subsequent event | any prior event)*.

We can illustrate the use of the probability square by applying it to the problem in the previous paragraph. Once the square is drawn the diagram is built up as follows (*see* Fig. 19.2):

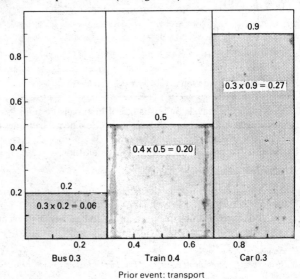

Figure 19.2   *Probability square*

(*a*) the base is divided pro rata to the probabilities of the method of travel – i.e. 0.3, 0.4, 0.3;

(*b*) each vertical rectangle is divided pro rata to the conditional probabilities of early and late arrival – i.e. 0.2, 0.5, 0.9;

(*c*) the area of each 'success' (arrive on time) is computed – i.e. 0.06, 0.20, 0.27;

(*d*) the areas are added – i.e. 0.06 + 0.20 + 0.27 = 0.53.

As can be seen, the sum of the success areas is the probability of a success – in this case, that my son arrives on time.

14. Bayes' Theorem. There is an interesting extension of the technique developed in **13** above, since where there is more than one possible prior event and the subsequent event actually proves to be a success, we can ask ourselves, 'what is the probability that a given possible prior event did, in fact, occur?'

Now note that this is *not* to ask simply what is the probability of a given prior event but what is the probability of a given prior event, *in view of the fact that the subsequent event occurred*. Thus, in the paragraph above, before my son arrives the probability of him using a given mode of transport are the 0.3, 0.4 and 0.3 previously detailed. However, *if he arrives on time* then we have an additional piece of information that changes the probabilities as to which mode he *actually* used.

Given, then, that he arrives on time, how can we compute the probability of each prior event having actually occurred? The answer to this lies in observing how frequently he would use each mode of transport on the successful occasions, i.e. on the occasions he arrives on time. (So that, for example, if he were always late when he walked, then if he arrived on time this very fact would have meant that the probability of him having walked must be 0 regardless of what the probability of him walking might have been before he set out.) Now as can be seen in Fig. 19.2 since there is a 0.06 probability of him travelling by bus *and* arriving on time then this event will occur 6 times in every 100 trips. Similarly, he will travel by train and arrive on time 20 times in every 100 and by car and arrive on time 27 times in every 100. And this means that out of the 53 times he actually arrives on time he will have travelled by bus 6 times, by train 20 times and by car 27 times. So the probability that, having arrived on time, he actually travelled by bus is 6/53, by train is 20/53 and by car 27/53. (Note, incidentally, that while before setting out he has only a 0.3 probability of having a lift with the neighbour, if he arrives on time he is more likely to have had a lift than not.)

This method can be put in formula terms as follows:

Given that a particular subsequent event S has actually occurred, then the probability that a given possible prior event E actually occurred is:

$$P(E|S) = \frac{P(E) \times P(S|E)}{P(S|\text{any prior event})}$$

Since $P(S|\text{any prior event})$ is computed by multiplying the probability of every E by the conditional probability of S given E occurred, and then adding the resulting products, $P(S|\text{any prior event})$ can be written as $\sum_{i=1}^{n}(P(E_i) \times P(S|E_i))$

$$\therefore P(E|S) = \frac{P(E) \times P(S|E)}{\sum_{i=1}^{n}(P(E_i) \times P(S|E_i))}$$

And this formula is known as *Bayes' Theorem*.

Example

There are three urns, one black containing 3 red and 7 white marbles, one red containing 4 yellow and 6 green marbles and one white containing 8 yellow and 2 green marbles. A marble is first drawn from the black urn. A second marble is then drawn from the urn having the same colour as the first marble drawn. If the second marble is yellow, what is the probability that the first marble was red?

Solution

Here the actual subsequent event is a yellow marble. There are only two sequential events which can give this result – the red–yellow and the white–yellow. The probability, then, of the second marble being yellow is:

Sequential event	Probability
R–Y (i.e. black and red urns)	$3/10 \times 4/10 = 12/100$
W–Y (i.e. black and white urns)	$7/10 \times 8/10 = 56/100$
$\therefore$ Probability of 2nd marble yellow =	$68/100$

So we can say P(1st marble red|2nd marble yellow)

$$= \frac{\text{Probability of R} \times P(Y|R)}{\text{Probability of 2nd marble Y}} = \frac{0.3 \times 0.4}{0.68} = 0.176$$

In other words, out of the 68 times (in 100) a yellow marble would be drawn, on only 12 occasions would a red marble have been drawn previously from the black urn. So the probability that the first marble drawn was red, given that the second marble was yellow, is $12/68 = 0.176$.

Simply using the formula above we would have:

$$P(\text{red}|\text{yellow}) = \frac{P(\text{red}) \times P(\text{yellow}|\text{red})}{P(\text{red}) \times P(\text{yellow}|\text{red}) + P(\text{white}) \times P(\text{yellow}|\text{white})}$$

$$= \frac{0.3 \times 0.4}{0.3 \times 0.4 + 0.7 \times 0.8} = 0.176$$

Progress test 19

(Answers in Appendix 4)

1. A games urn contains 4 white and 1 black marble. A player takes out the marbles at random one by one. Every time he takes out a white marble he is paid £1. When he takes out the black marble his turn is over. What is his expectation?

2. The six faces of a Crown and Anchor die show the four card suits, a crown and an anchor. The board is divided into six squares, each showing one of these six symbols. Players place their stakes on these squares and the banker then throws three dice. Each player compares his stake symbol with the three upturned die faces. If one face is the same as his staked symbol he receives his stake back and as much again. If two faces are the same he receives his stake back and twice as much again, while if all three faces are the same he receives his stake back and three times as much again. What is the *banker's* expectation from this game?

3. There are two urns, A and B. In A are 2 black and 1 white marbles and in B 1 black and 2 white. A marble is chosen at random from A and placed in B. A marble is then chosen at random from B. If this marble is white a prize of 60p is won, but if it is black a penalty of 60p has to be paid. What is the financial expectation of any hopeful potential prize-winner?

4. A purse contains ten 1p coins, six 2p pieces and four 5p pieces. In a random draw of three coins what is the probability of drawing a total value of:

   (a) exactly 8p; and
   (b) less than 7p?

(CA)

5. The probability of my arriving at work less than 5 minutes late is 0.7 and of arriving 5 or more minutes late is 0.2. If I am less than 5

minutes late the probability of a reprimand is 0.4 while if I am 5 or more minutes late it is 0.9.

(a) What is the probability that I will avoid a reprimand tomorrow?

(b) If I got a reprimand today, what is the probability that I was less than 5 minutes late?

6. The probability that my son will use my car while I am at work is 0.3. The probability that my monthly garage bill will be reasonable if he does not use my car is 0.2, while if he does it will be 0. The bill has been unreasonable for the past three months. What is the probability that my son has, on occasions, used my car during this period?

7. Poorkwality Ltd. manufacture poorkwality units on three machines, A, B and C. A can manufacture 300 units an hour and produces 10 per cent substandard, B can manufacture 200 units per hour and produces 15 per cent substandard while C can only manufacture 100 units per hour and produces 24 per cent substandard. Incoming parts are allocated to machines on a basis of machine availability. What is the probability that:

(a) an individual unit is produced on machine A and emerges substandard?

(b) an individual unit selected at random from the finished output proves to be substandard?

(c) a substandard unit found by chance in the finished output was manufactured on machine A?

8. (a) A glass bottle manufacturer has three inspection points – one for size, the second for colour and the third for flaws such as cracks and bubbles in the glass. The probability that each inspection point will incorrectly accept or reject a bottle is 0.02.

What is the probability that:

(i) a perfect bottle will be passed through all inspection points;

(ii) a bottle faulty in colour and with a crack will be passed;

(iii) a bottle faulty in size will be passed.

(b) Two machines produce the same type of product. The older machine produces 35 per cent of the total output but eight in every hundred are normally defective. The newer machine produces 65 per cent of the total output and two in every hundred are defective.

Determine the probability that a defective product picked at random was produced by the older machine.

(CIMA)

9. If 2 per cent of the population suffer from an eye defect which is difficult to diagnose and a test is discovered which, while identifying with certainty as a result of test failure those people who have the defect, carries a 5 per cent probability that someone not suffering from the defect will fail the test, what is the probability that an individual failing the test will, in fact, suffer from the defect?

10. (*a*) State briefly whether the following pairs of events are independent, and why.

(*i*) Winning two successive prizes in the monthly Premium Bond draws.

(*ii*) Earning a large salary and paying a large amount of income tax.

(*iii*) Being drunk while driving and having an accident.

(*iv*) Being an accountant and having large feet.

(*v*) Any two mutually exclusive events.

(*b*) A manufacturer assembles a toy from four independently produced components, each of which has a probability of 0.01 of being defective. What is the probability of a toy being defective?

(*c*) A factory has a machine shop in which three machines (A, B and C) produce 100cm aluminium tubes. An inspector is equally likely to sample tubes from A and B, and three times as likely to select tubes from C as he is from B. The defective rates from the three machines are:

$$A \quad 10\% \quad B \quad 10\% \quad C \quad 20\%$$

What is the probability that a tube selected by the inspector:

(*i*) is from machine A:

(*ii*) is defective;

(*iii*) comes from machine A, given that it is defective?

(CIMA)

11. A coin has been tossed ten times and has fallen 'heads' each time. State the three arguments which conclude, validly or otherwise, that on the next toss the coin is more likely, equally likely and less

260 Sampling theory and inference

likely to fall 'tails'. Which argument do you support and why?

Assignment

There are three pairs of dice with faces marked 1 to 6, 2 to 7, and 3 to 8 respectively. A 'banker' throws one of the first pair of dice. If he throws a 1 he then throws both dice in the first pair. If he throws a 2 or a 3 then he throws the second pair of dice. If he throws a 4, 5 or 6 then he throws the third pair of dice. A player behind a screen is then told what the total value of the pair of thrown dice is, and he has to announce how the very first single die fell. Find for each possible score that he can be told the probabilities that the first die fell 1, 2, 3, 4, 5 or 6 (e.g. if he is told the score is 16 then he can be sure the third pair of dice were thrown and hence the probability that the first die fell 1, 2 or 3 is zero and the probability that it fell 4, 5 or 6 is one third for each value).

20
Probability distributions

In this chapter we look at the concept of a probability distribution as well as two particular distributions, the binomial and the Poisson. In the next chapter we will look at a third distribution, the normal curve. In this chapter we also look at the mathematics of combinations since this is fundamental to understanding probability distributions.

Introduction

First, it is necessary to understand what is meant by a probability distribution.

1. Random variable. Often events cannot be quantified, i.e. given numerical values. One is either struck by lightning or one is not. A white marble can be drawn from an urn or a red or a black marble but such events have no numerical value. On the other hand a throw of a die can be quantified – the event simply has the value of the face showing. Where random events can be quantified then the values that can arise are termed random variables. Formally a *random variable* is the *quantified value of a random event*. Another way of looking at a random variable is to see it as any value that depends on chance.

Although it is possible to create random variables where marbles are drawn from an urn – by, say, valuing a white at 1, a red at 2 and a black at 3 – in this chapter we will be looking only at what might be called 'natural' random variables. Indeed, we will only consider random variables which are numbers that can be counted, e.g. number of coins falling 'heads' or number of red marbles drawn from an urn. As long as such counted numbers arise by chance they are all random variables.

2. Probability distributions. We have seen that a frequency distribution shows us how many times given values in a range of values occur. A probability distribution is very similar – it shows us how probable given random variable values in a range of such values are.

Let us look at a very simple example. If we toss two coins we can obtain 0, 1 or 2 'heads'. If we prepare a table showing the probabilities of all the random variable values we will have a probability distribution:

Number of 'heads'	Sequential event		Probability	
0	T	T	0.5×0.5	= .25
1	$\begin{cases} \text{H} \\ \text{T} \end{cases}$	$\begin{matrix} \text{T} \\ \text{H} \end{matrix}$	$\left. \begin{matrix} 0.5 \times 0.5 \\ 0.5 \times 0.5 \end{matrix} \right\}$	= .50
2	H	H	0.5×0.5	= .25
				1.00

Note that, just as in a frequency distribution the sum of the distribution frequencies was the total of all the occurrences, so the sum of a probability distribution must equal 1 – otherwise one or more probabilities have been left out of the distribution or computed wrongly.

3. Populations and samples. Generally speaking probability distributions are used to estimate the probability of obtaining a particular result if a sample is taken from a population with a known composition. Thus, in the previous paragraph our two-coin toss was 'taken' from the infinite population of all the possible two-coin tosses – which we subjectively know will have the same number of 'heads' as 'tails' (so P(Head)=0.5 and P(Tail)=0.5). From the probability distribution we computed we can say that if we took a sample of, say, 100 two-coin tosses our estimate of the number of zero 'heads' would be $100 \times 0.25 = 25$, the number of single 'heads' would be $100 \times 0.5 = 50$, and the number of two 'heads' would be $100 \times 0.25 = 25$.

In this chapter it will be assumed the population composition is known and the probability of a given composition of a sample is to be estimated. Later, in the chapter on sampling theory, we will deduce facts about the population from the sample.

4. Constant probabilities. When looking at conditional probability we saw that the probability of selecting a particular coloured marble from an urn changed as the number of marbles in the urn decreased. Whenever a sample item is taken from a population the probabilities change for the same reason. However, where populations are large this change is so small as to be insignificant. This means that where populations are large the probability of selecting a particular item remains constant regardless of how many items have already been selected.

For the rest of Part 6 this assumption of a constant probability will be made.

5. Symbols. In this chapter the following two symbols will be added to our other probability symbols:

(*a*) *n*. This will be the sample size, i.e. the number of items selected from the population.

(*b*) P(*x*). This will symbolise the probability that exactly *x* items out of the *n* selected at random will have a defined characteristic (e.g. if 'head' is a defined characteristic, *n* is 20 and *x* is 12 then P(*x*) is the probability that 12 of the 20 sample items will be 'heads').

Combinations

Before we can turn to probability distribution theory we must look first at the mathematics underlying combinations. However, as this topic is essentially a mathematical topic rather than a statistical one, the treatment here will necessarily be brief. If at the end the development of the final formula is still not clear to the reader he should either take it on trust or turn to a mathematical text for further clarification.

6. !. There is a mathematical symbol ! (spoken 'factorial') that means 'multiply together all the numbers in a countdown from the number preceding the !'. So $7! = 7 \times 6 \times 5 \times 4 \times 3 \times 2 \times 1$. And in this chapter *n*! means multiply all the numbers in a countdown from the total number in the sample.

Note, incidentally that 0! is deemed to be 1, i.e. $0! = 1$.

7. Number of permutations of n items taking x at a time.
Imagine we have seven men – A, B, C, D, E, F and G – in a waiting room and we intend to photograph them three at a time sitting on a chair, a stool and a box. How many different photographs can we take?

The answer is, quite a number. First of all, we can have each of seven men sitting in turn in the chair. As each sits there we can partner him with any one of the remaining six men sitting on the stool. So we have 7×6 partnerships. And each of these partnerships can team up with any one of the remaining five men in the waiting room sitting on the box. So that means 7×6 partnerships with five team-mates each gives $7 \times 6 \times 5$ (=210) different photographs.

Reflecting on how this total was found it will be seen that if the total men in the waiting room is n, then we have the first three terms of $n!$ (i.e. the first three terms of $7! = 7 \times 6 \times 5$). Now, mathematically we can cut off the tail of a factorial by dividing by a smaller factorial. Take, for instance, cutting off the tail of $7!$ We can do this as follows:

$$7 \times 6 \times 5 = \frac{7 \times 6 \times 5 \times 4 \times 3 \times 2 \times 1}{4 \times 3 \times 2 \times 1}$$

Since the last four terms of the denominator and the numerator cancel out we have:

$$7 \times 6 \times 5 = 7!/4! = n!/4!$$

This is fine as far as it goes, but how would we know in a different situation the size of the smaller factorial? Well, since the head of the factorial ($7 \times 6 \times 5$) has as many terms as men being photographed, the smaller factorial must be the number of men left in the waiting room when we are actually taking a photograph. And if we call the number of men whose photograph is being taken x, then the smaller factorial is $n - x$. So:

Number of different photographs (permutations) of n men (items) when taking x men at a time $= n!/(n-x)!$

Examples

(a) Number of permutations of 7 items taking 3 at a time $= 7!/(7-3)! = 7!/4! = 7 \times 6 \times 5$

(*b*) Number of permutations of 15 items taking 5 at a time = $15!/(15-5)! = 15!/10! = 15 \times 14 \times 13 \times 12 \times 11$.

8. Number of combinations of n items taking x at a time – $\binom{n}{x}$.

Now it may be that somebody looking at your collection of photographs complains that since the same three men will appear in many photographs (e.g. (A on chair, B on stool, C on box) and (B on chair, A on stool, C on box)) they are not really different at all. Your collection, taking into account *different orders* as well as different men is, in fact, the complete set of *permutations*. His view, which is to disregard order and only define 'different' as different groupings of men, means he is only concerned with a number of *combinations*. And the next question is, if he wants different combinations only, how many photographs must you show him?

This is not too difficult. Each group of three men can be arranged as follows: each of the three men can sit on the chair in turn. This leaves two others who can sit on the stool – i.e. 3×2 partnerships. Since now only one man is left he must perforce sit on the box. So each group of three men gives rise to $3 \times 2 \times 1 = 3!$ photographs. This means you need to show him only one photograph in $3! = 6$ – i.e. the number of permutations of 7 items taking three at a time divided by $3!$

$$= \frac{(7!/(7-3)!)}{3!} = \frac{7!}{(7-3)!3!}$$

Since the final division by $3!$ is in fact $x!$ – the number of items taken at a time – we can generalise from this argument and say that:

Number of *combinations* of n items taking x at a time $= \dfrac{n!}{(n-x)!x!}$

And this complete formula can be symbolised as either $\binom{n}{x}$ or $^{n}C_{x}$

Example

(*a*) Number of combinations of 5 items taking 3 at a time =

$$\binom{5}{3} = \frac{5!}{(5-3)!3!} = \frac{5 \times 4 \times 3 \times 2 \times 1}{(2 \times 1) \times (3 \times 2 \times 1)} = 10$$

i.e. if we have items A, B, C, D and E we have combinations ABC, ABD, ABE, ACD, ACE, ADE, BCD, BCE, BDE, CDE.

(*b*) Number of combinations of 15 items taking 5 at a time =

$$\binom{15}{5} = \frac{15!}{(15-5)!5!}$$

$$= \frac{15 \times 14 \times 13 \times 12 \times 11 \times 10 \times 9 \times 8 \times 7 \times 6 \times 5 \times 4 \times 3 \times 2 \times 1}{(10 \times 9 \times 8 \times 7 \times 6 \times 5 \times 4 \times 3 \times 2 \times 1) \times (5 \times 4 \times 3 \times 2 \times 1)}$$

$$= \frac{15 \times 14 \times 13 \times 12 \times 11}{5 \times 4 \times 3 \times 2 \times 1}$$

Note that a shorter way of writing out this last calculation is to cancel out factorials, i.e.

$$\frac{15!}{(15-5)!5!} = \frac{15 \times 14 \times 13 \times 12 \times 11 \times 10!}{10! \times 5!}$$

Now the two 10!s cancel leaving $\dfrac{15 \times 14 \times 13 \times 12 \times 11}{5 \times 4 \times 3 \times 2 \times 1}$

The binomial distribution

Imagine we have an urn containing 100,000 marbles, 40,000 of which are red. A random sample of 4 marbles is taken. What are the probabilities that in this sample there are 0 red marbles, 1 red marble, 2 red marbles, 3 red marbles and 4 red marbles respectively (i.e. what probabilities will the random variable values 0, 1, 2, 3 and 4 take)?

9. Sequential event analyses. In this problem we first note that p, the constant probability of a success, is $40,000/100,000 = 0.4$. So q, the probability of a failure, is $1 - 0.4 = 0.6$.

Now we can prepare a sequential event analysis in the normal way, noting that the first column – the number of red marbles in the sample – can be looked at as a list of the random variable values or a list of compound events:

Number of red marbles – x	Sequential event*		Probability
0	– – – –	$0.6 \times 0.6 \times 0.6 \times 0.6$	$0.6^4 = 0.1296$
1	R – – –	$0.4 \times 0.6 \times 0.6 \times 0.6$	
	– R – –	$0.6 \times 0.4 \times 0.6 \times 0.6$	$4 \times 0.4 \times 0.6 \times 0.6 \times 0.6 = 0.3456$
	– – R –	$0.6 \times 0.6 \times 0.4 \times 0.6$	
	– – – R	$0.6 \times 0.6 \times 0.6 \times 0.4$	

*Non-red marbles symbolised as '–'

2	R R - -	$0.4 \times 0.4 \times 0.6 \times 0.6$	
	R - R -	$0.4 \times 0.6 \times 0.4 \times 0.6$	
	R - - R	$0.4 \times 0.6 \times 0.6 \times 0.4$	$6 \times 0.4 \times 0.4 \times 0.6 \times 0.6 = 0.3456$
	- R R -	$0.6 \times 0.4 \times 0.4 \times 0.6$	
	- R - R	$0.6 \times 0.4 \times 0.6 \times 0.4$	
	- - R R	$0.6 \times 0.6 \times 0.4 \times 0.4$	
3	R R R -	$0.4 \times 0.4 \times 0.4 \times 0.6$	
	R R - R	$0.4 \times 0.4 \times 0.6 \times 0.4$	$4 \times 0.4 \times 0.4 \times 0.4 \times 0.6 = 0.1536$
	R - R R	$0.4 \times 0.6 \times 0.4 \times 0.4$	
	- R R R	$0.6 \times 0.4 \times 0.4 \times 0.4$	
4	R R R R	$0.4 \times 0.4 \times 0.4 \times 0.4$	$0.4^4 = \underline{0.0256}$
			1.0000

And this is the probability distribution of a random sample of four items taken from a large population where P(red) = 0.4.

10. Binomial distribution. If the analysis in the previous paragraph is studied the following points will be observed:

(*a*) There is considerable repetition of the probabilities relating to the sequential events. If these are examined it will be seen that:

(*i*) 0 reds involves q to the power of a sample size;

(*ii*) 1 red involves q to the power of (sample size − 1) and p to the power of 1;

(*iii*) 2 reds involves q to the power of (sample size − 2) and p to the power of 2;

(*iv*) 3 reds involves q to the power of (sample size − 3) and p to the power of 3;

(*v*) 4 reds involves p to the power of 4.

So if x is the number of reds we can say that the probability of x is in some way related to $p^x q^{n-x}$. (Note: p^0 and p^0 are both equal to 1.)

(*b*) The number by which we must multiply $p^x q^{n-x}$ is the number of combinations of n items taking x at a time. If, for instance, you take the probability of 2 reds, then looking at the analysis you can see that the number you multiply $p^x q^{n-x}$ by is the combination of 4 items taking 2 at a time.

Our analysis, as it happens, illustrates the fact that where we have only two possible different characteristics in a population then in a sample of n items the probability of $0, 1, 2, 3 \ldots$, i.e. x items, having a given characteristic is given by the formula:

$$P(x) = \binom{n}{x} p^x q^{n-x}$$

Applying this formula for all the random variable values between $x=0$ and $x=n$ gives us the *binomial distribution*.

11. Illustration of the binomial distribution. Let us now apply this formula to our illustrative problem – i.e. where $p=0.4$, $q=0.6$ and $n=4$. Here the formula becomes:

$$\binom{4}{x} 0.4^x 0.6^{4-x}$$

or, since (*see* **8**):

$$\binom{4}{x} = \frac{4 \times 3 \times 2 \times 1}{(4-x)! x!} = \frac{24}{(4-x)! x!}$$

Then

$$P(x) = \frac{24}{(4-x)! x!} \times 0.4^x \times 0.6^{n-x}$$

Preparing the binomial distribution layout we obtain:

x (no. red marbles)	$P(x)$ (*i.e. probability of* x *marbles in sample*)
0	$\dfrac{24}{(4 \times 3 \times 2 \times 1)! \times 0!^*} \times 0.4^{0^*} \times 0.6^{4-0} = 1 \times 1 \times 0.6^4 = 0.1296$
1	$\dfrac{24}{(3 \times 2 \times 1)! \times 1!} \times 0.4^1 \times 0.6^{4-1} = 4 \times 0.4 \times 0.6^3 = 0.3456$
2	$\dfrac{24}{(2 \times 1)! \times (2 \times 1)!} \times 0.4^2 \times 0.6^{4-2} = 6 \times 0.4^2 \times 0.6^2 = 0.3456$
3	$\dfrac{24}{1! \times (3 \times 2 \times 1)!} \times 0.4^3 \times 0.6^{4-3} = 4 \times 0.4^3 \times 0.6^1 = 0.1536$
4	$\dfrac{24}{0! \times (4 \times 3 \times 2 \times 1)!} \times 0.4^4 \times 0.6^{4-4} = 1 \times 0.4^4 \times 0.6^{0^*} = 0.0256$
	1.0000

* Remember $0! = 1$ and any number to the power of $0 = 1$.

This table gives the same result as before and so confirms the binomial distribution formula.

12. Application of the binomial distribution. We have now looked at the binomial distribution formula and seen how that formula can be applied. At this point it should be appreciated that the validity of the formula depends very much on the fact that to all intents and purposes p remains virtually unchanged during the entire selection of samples. To the extent that this is not true the formula will only give us an approximation of the true probability. Note, however, that where the population is infinite p never changes and in such an instance the formula gives us the exact probability. Infinite populations exist where there is coin tossing or dice throwing for, as we have seen, any individual trial is merely one of an infinite number of potential trials.

13. Mean and standard deviation of the binomial distribution. It is stated without proof (though the student can test the formula on actual distributions if he wishes) that the mean and standard deviation of a binomial distribution are:

Mean $= np$
Standard deviation $= \sqrt{npq}$

Example
Mean of distribution in **11** $= 4 \times 0.4 = 1.6$ red marbles, i.e. if thousands of samples of 4 marbles were taken, then the average number of red marbles per sample would be 1.6.

Standard deviation of distribution in **11** $= \sqrt{(4 \times 0.4 \times 0.6)} = 0.980$ red marbles.

The Poisson distribution

Next let us imagine that in our 100,000-marble urn there are now only 120 red marbles but that our sample size is to be 1,000. What now are the probabilities of having 0, 1, 2, 3, 4 etc. red marbles in our random sample?

In this instance p is obviously only $120/100,000 = 0.0012$ and so employing the binomial distribution will involve using the formula

$$P(x) = \frac{1,000!}{(1,000-x)!x!} \times 0.0012^x \times 0.9988^{1,000-x}$$

where x takes every value from 0 to 1,000. In this situation computing only $P(x=0)$ means that we will have to find $0.9988^{1,000}$, while computing $P(x=2)$ requires us to find $\frac{1,000!}{998!\,2!} \times 0.0012^2 \times 0.9988^{998}$. Clearly, a short-cut method would be much appreciated in this situation.

14. The Poisson distribution. Fortunately such a short cut exists since in these circumstances the *Poisson distribution* gives us a very good approximation to the binomial distribution. The formula for this distribution is:

$$P(x) = e^{-a} \times \frac{a^x}{x!}$$

where a = average number of items per sample.

At first sight this looks rather complicated since we still have to find a and then compute e^{-a}, whatever e may happen to be. However, these problems are trivial. It turns out that a can be found from the formula $a = np$, and that e is a constant (rather like π) which means that a table for e^{-a} can be prepared so that finding e^{-a} for any value of a simply involves looking up the table – and such a table is given in Appendix 3.

Example

Taking our particular illustration, $p = 0.0012$ and $n = 1,000$, so $a = 0.0012 \times 1,000 = 1.2$. And Appendix 3 shows us that $e^{-1.2}$ (i.e. e^{-a}) = 0.3012.

15. An illustrative Poisson distribution. Let us now apply the Poisson distribution to the specific case where a 1,000-marble sample is selected from a large population in which the probability of selecting a red marble is 0.0012. As we saw in **14** above, in this case $a = 1.2$ and $e^{-a} = 0.3012$. This means that our Poisson distribution formula becomes:

$$P(x) = 0.3012 \times \frac{1.2^x}{x!}$$

Since we want the probability of having 0, 1, 2, 3, 4, etc., red marbles in our sample of 1,000 the layout we need is as follows:

x (No. red marbles)	P(x) (i.e. probability of x red marbles in sample)
0	$0.3012 \times \dfrac{1.2^0}{0!} = 0.3012$
1	$0.3012 \times \dfrac{1.2^1}{1!} = 0.3614$
2	$0.3012 \times \dfrac{1.2^2}{2!} = 0.2169$
3	$0.3012 \times \dfrac{1.2^3}{3!} = 0.0867$
4	$0.3012 \times \dfrac{1.2^4}{4!} = 0.0260$
5	$0.3012 \times \dfrac{1.2^5}{5!} = 0.0062$
6 . . . 1,000	Since we know that the whole probability distribution must add up to 1 (for it is certain there will be 0 or 1 or 2 or 3 or 4 . . . or 1,000 red marbles in the sample) all these P(x)s must *collectively* be the balance required to make the total equal 1 = 0.0016
	1.0000

As can be seen, the probabilities drop rapidly once x passes 2, and that in fact the probability of a sample of 1,000 containing more than 5 reds (let alone a specific number of reds) is a mere 0.0016.

16. When a Poisson distribution can be used. Again, the Poisson distribution can only be used in particular circumstances (though those circumstances are very common). It can, in fact, be used *whenever the sample size is very large and the probability of a success is very small*, i.e. when n is very large and when p is very small. This wording of this restriction, of course, raises the problem of how large is very large and how small is very small. As was said in **14**, the Poisson distribution is an approximation to the binomial distribution and so the larger n and the smaller p the better the approximation will be. However, for an all-round rule the Poisson distribution can usually be used when p is less than 0.1 (when a sample size of less than 20 would in practice be hardly worth taking).

17. Mean and standard deviation of the Poisson distribution. Again it is stated, without proof, that the mean and standard deviation of the Poisson distribution are:

Mean $= a$
Standard deviation $= \sqrt{a}$

Example

Taking the distribution in **15**:

Mean $= 1.2$ red marbles per sample
Standard deviation $= \sqrt{1.2} = 1.095$ red marbles

18. The Poisson distribution and an unknown sample size. The Poisson distribution has one very interesting (and useful) feature – it can be used even though the sample size is not known. This is because the Poisson formula only requires the *average* number of successes to be known and we can know the average without knowing either n or p. For example, to take the classical case, assume that we are studying thunderstorms and have been counting the number of lightning flashes. We find that over a total period of 100 minutes there were 120 flashes, i.e. an *average* of 1.2 flashes a minute. Now if we regard a particular minute as a very large sample of many very small moments of time and if there is a very small but constant probability of a single lightning flash occurring during each moment of time, then we can regard the pattern of the flashes as forming a Poisson distribution in which $a = 1.2$. So we can prepare a Poisson probability distribution that will tell us the probability of having in a given minute 0, 1, 2, 3, 4 etc. lightning flashes. (Since a here is identical to the a in **15** the probability distribution is exactly the same as that in **15** – i.e. for 'red marbles' in the layout you can substitute 'lightning flashes'.)

19. Graphing a probability distribution. As with other statistical data, further insight into a probability distribution situation can be obtained by graphing the distribution. This is done by plotting the random variable along the horizontal axis and probabilities along the vertical axis. The distribution is then graphed by constructing a

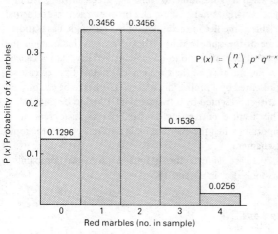

$$P(x) = \binom{n}{x} p^x q^{n-x}$$

(a) Binominal distribution $n = 4$ $p = 0.4$ $q = 0.6$

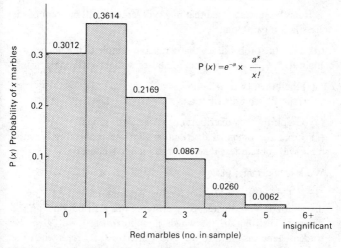

$$P(x) = e^{-a} \times \frac{a^x}{x!}$$

(b) Poisson distribution $a = 1.2$

Figure 20.1   *Binomial and Poisson distributions*

vertical rectangle over each random variable value to a height equal to the probability of that value.

Note that when it is desirable to show the graph in the form of a histogram the rectangles must adjoin and be drawn so that the actual random variable value lies in the middle of the rectangle base (and all bases must be of the same width). When drawn in this manner the combined probability of a group of random variable values can be found by adding the areas of their rectangles (though first, of course, the area represented by a probability of, say, 0.01 has to be established just as the area represented by a frequency of 1 had to be established in 8:3). This feature of treating probabilities as histogram areas becomes important when the distribution is a normal curve – the topic of the next chapter.

In Fig. 20.1 are shown the graphs of the two probability distributions looked at in **11** and **15**.

Progress test 20

(Answers in Appendix 4)

1. It has been established that in a very large city 10 per cent of the houses lack a bathroom.

(*a*) Find the probabilities that a random sample of ten houses will contain 0, 1, 2, 3, 4, 5, 6 or more houses without bathrooms, using:

   (*i*) the binomial distribution;
   (*ii*) the Poisson distribution.

(*b*) Graph the two distributions on the same graph.
(*c*) From the binomial distribution find the probability that the sample will contain less than two bathroomless houses.

Work to 4 decimal places.

Note: This is very much a borderline situation for using a Poisson distribution. Nevertheless, comparison of this distribution with the binomial distribution (to which it approximates) will give the student some insight into the extent of error that arises when the Poisson distribution is used in lieu of the binomial distribution.

2. Bus-Hire Limited has two coaches which it hires out for local use

by the day. The number of demands for a coach on each day is distributed as a Poisson distribution, with a mean of two demands.

(a) On what proportion of days is neither coach used?

(b) On what proportion of days is at least one demand refused?

(c) If each coach is used an equal amount, on what proportion of days is *one* particular coach not in use?

(CIMA)

3. An airline deliberately overbooks its local Mini flights because it knows from experience that not all passengers who book for a given flight actually arrive for that flight.

It is assumed that the probability of any booked passenger arriving for a given flight is 0.8; this is independent of the probability of any other passenger arriving.

The airline takes ten bookings for an 8-seater aircraft. Use the binomial distribution to find the probability for a given flight:

(a) that the aircraft takes off full;

(b) that the aircraft takes off with at least two empty seats.

(CIMA)

Assignments

1. Toss ten coins and count the number of 'heads'. Repeat 100 to 200 times and check your results against the binomial distribution values.

2. Draw a grid of 1cm squares on a large piece of paper. Then throw a dart at random repeatedly at the paper. After 200 to 300 throws count the number of holes in each square and test these numbers against the Poisson distribution values. Is there anything in this experiment which militates against your results forming a Poisson distribution?

21
The Normal curve

Of all the probability distributions the most important is the one that gives us the Normal curve.

1. Probability distribution of tossing 16 coins. Imagine a trial consists of tossing 16 coins. We wish to know the probability of obtaining 0 'heads', 1 'head', 2 'heads', 3 'heads', etc. Here we will clearly use the binomial distribution where n is 16, p is 0.5 and q is 0.5. So, where x is the number of 'heads':

$$P(x) = \binom{16}{x} \times 0.5^x \times 0.5^{16-x} = \binom{16}{x} \times 0.5^{16}$$

If we apply this formula to every possible value of x we obtain the following distribution (the cumulative probabilities are needed for later reference):

x	$P(x)$	Cum. Prob.	x	$P(x)$	Cum. Prob.	x	$P(x)$	Cum. Prob.
0	*	*	6	0.1222	0.2272	12	0.0278	0.9895
1	0.0002	0.0002	7	0.1746	0.4018	13	0.0085	0.9980
2	0.0018	0.0020	8	0.1964	0.5982	14	0.0018	0.9998
3	0.0085	0.0105	9	0.1746	0.7728	15	0.0002	1.0000
4	0.0278	0.0383	10	0.1222	0.8950	16	*	1.0000
5	0.0667	0.1050	11	0.0667	0.9617			

*Insignificant

The graph of the distribution with its frequency curve superimposed is shown in Fig. 21.1.

Note the following points.

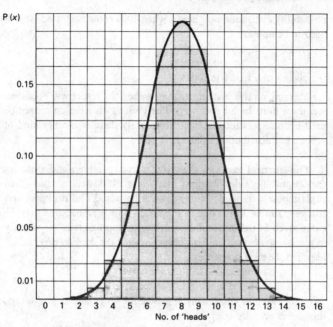

P (x)

0.15

0.10

0.05

0.01

0 1 2 3 4 5 6 7 8 9 10 11 12 13 14 15 16
No. of 'heads'

Figure 21.1 *Binomial distribution*

(a) The probability of any range of random variable values can be found merely by adding the individual probabilities. So:

P(x<6)=0.1050 (cumulative probabilities 0 to 5)
P(9 or 10 or 11 'heads')=0.1746+0.1222+0.0667=0.3635

Note also that P(9 or 10 or 11 'heads')=difference between two cumulative probabilities=0.9617−0.5982=0.3635.

(b) Just as in a histogram of frequencies, the area of each rectangle in Fig. 21.1 is proportional to the probability of the variable value. And if the area of the whole histogram is taken as 1, the area of each rectangle will measure the probability of obtaining that variable if 16 coins are tossed.

2. Probability distribution of tossing 1000 coins. If we want the probability distribution of 'heads' when 1000 coins are tossed we will need to compute:

$$\binom{1000}{x} \times 0.5^{1000} = \frac{1000!}{(1000-x)!x!} \times 0.5^{1000}$$

where x runs from 0 to 1000.

Reflecting on this mathematical marathon the reader will doubtless feel there must be a better way. The Poisson distribution cannot be used because p is too large but fortunately there is another kind of probability distribution available to us.

3. The Normal curve. *If* the reader were to persevere and construct the probability distribution graph for 1000 coins and then superimpose a frequency curve upon it he would find his curve was still very similar to the one in Fig. 21.1. Now it so happens that the curve given by this probability distribution approximates very closely to a *mathematical* curve. This curve is called the *Normal curve* and it has the following properties:

(*a*) it is a symmetrical, bell-shaped curve;

(*b*) its mean lies directly below its peak;

(*c*) the two tails continually approach the horizontal axis, although they never actually reach it;

(*d*) its formula, for those who, out of curiosity, would like to know it, is:

$$y = \frac{1}{\sigma\sqrt{2\pi}} e^{-\frac{1}{2}\left(\frac{x-\mu}{\sigma}\right)^2}$$

4. Areas of the Normal curve. Because the Normal curve is a mathematical curve it is possible to find the areas under the different parts of it by integration. This is very important for it means that if our probability distribution forms a Normal curve – and many probability distributions do – then we can find the probability of a given range of random variable values occurring (e.g. probability of obtaining between 300 and 400 'heads' in 1000-coin toss) by simply finding the area under the Normal curve above this range on the horizontal axis as a proportion of the total Normal curve area. And this is far easier than computing the binomial probabilities.

5. The standard Normal distribution. At this point the reader may feel he would rather compute the binomial probabilities than try and integrate the expression in **3**. Fortunately, though, we ourselves do not have to do any integrating for every Normal curve we have can in practice be fitted to a standard curve for which all the integration has been done. This standard curve is called the *standard Normal distribution* and is a Normal curve having the following features (*see* Fig. 21.2):

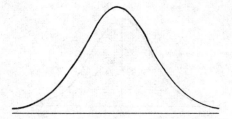

Figure 21.2 *Normal curve of distribution*

(*a*) its mean is zero;
(*b*) its standard deviation (σ) is 1;
(*c*) its total area is 1.

And all the different areas under this curve have been computed and published as tables (e.g. *see* Appendix 3.**1**).

To illustrate three of these areas Fig. 21.3 shows the area under the standard Normal distribution curve for 1σ, 2σ and 3σ. Note that to all intents and purposes the whole of the Normal curve area lies between -3 and $+3$ standard deviations, and that 19/20ths of the area lies between -2σ and $+2\sigma$.

6. Converting an actual curve to standard. So far so good. However, our probability distribution curve does not have a mean of 0 and a standard deviation of 1, so how can we use the tables relating to the standard Normal distribution?

The answer is that we tailor our curve so that it just fits over the standard curve, i.e. we arrange for the mean of our curve to lie at zero on the standard curve and for our curve standard deviations to fit the standard deviations of the standard curve. To do this we have to find

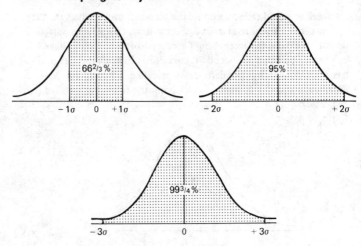

Figure 21.3 *Approximate areas beneath the Normal curve*

the mean and standard deviation of our probability distribution, but this is easily done using the formula in 20:**13**. So:

Mean = np = 1000 × 0.5 = 500 'heads'
Standard deviation = $\sqrt{npq}$ = $\sqrt{1000 \times 0.5 \times 0.5}$ = 15.81 'heads'

Our mean of 500 'heads', then, lies at zero on the standard distribution and the 1σ point above the mean on the standard curve is the 515.81 'heads' on our actual distribution (*see* Fig. 21.4).

7. z. Normal curve tables normally show the areas lying under the curve between the mean of the standard Normal distribution and any standard deviation value – e.g. the area under the curve between the mean and the 1σ value is approximately one-third (half the two-thirds shown for the area between the 1σ values either side of the mean). So, to use the tables we must convert our values into the standard values. This simply involves starting at the mean of our distribution (500) and then measuring off our value along the horizontal axis in standard deviations (i.e. in units of 15.81). This measurement is symbolised as z: z, then, tells us in units of standard deviation how far from the mean a given value lies along the horizontal axis. Clearly, to find z from an actual distribution value, say x, you simply find how far x is from the

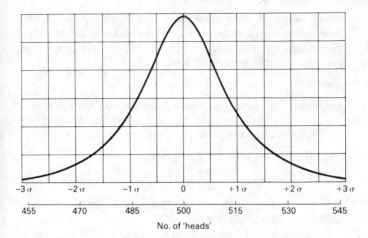

Figure 21.4 *Coin-tossing distribution converted to standard Normal curve*

mean and then divide this distance by the distribution standard deviation. In other words:

$$z = \frac{x - \mu}{\sigma}$$

where μ and σ are the mean and standard deviation of the actual distribution.

Example

Find in respect of our probability distribution illustration z when x is 530 'heads'.

$z = \dfrac{530 - 500}{15.81}$ which is approximately $+2$

So 530 'heads' is equivalent to 2σ on the standard Normal distribution curve.

8. Using the Normal curve table. Having found z, the next step is to look at the Normal curve table (Appendix 3) and see what area is associated with that value of z. Thus, if z were 1.8, the Appendix shows us the area would be 0.4641, i.e. that the area under the curve from the mean up to 1.8σ is 0.4641 of the total area under the curve. Note, incidentally, that if we want the area under a curve between two values, neither of which is the mean, then we simply find the two areas under the curve between the mean and each of the values respectively, and then combine – by addition or subtraction – these two areas to obtain the area we want.

Example

Knowing the mean and standard deviation of our illustrative probability distribution are 500 and 15.81 'heads', find the following areas (z to be rounded to one decimal place):

(a) Between the mean and 460 'heads':

For 460 'heads', $z = \dfrac{460 - 500}{15.81} = -2.5$*

So area between mean and 460 'heads' is 0.4938, i.e. the area associated with $z = 2.5$ in Appendix 3.

*the negative value here simply indicates that x is below the mean.

(b) Between 460 and 520 'heads':

For 460 'heads', $z = -2.5$ (*see* (a))

So area between 460 and mean = 0.4938

For 520 'heads', $z = \dfrac{520 - 500}{15.81} = +1.3$

So area between 520 and mean = 0.4032.

So since the two values lie either side of the mean the required area is the sum of the two areas = 0.8970.

(c) Between 510 and 520 'heads':

For 510 'heads', $z = \dfrac{510 - 500}{15.81} = 0.6$

So area between 510 and mean = 0.2257.

And area between 520 and mean (*see* (b)) = 0.4032.

And since the two values lie on the same side of the mean the required area is the difference between the two areas = 0.1775.

Note: Giving the areas to four decimal places when z has been rounded to one decimal place results in spurious accuracy. However, the figures have been left as they are so that the reader can see how the problems were solved.

9.   Areas and probabilities. It was said earlier that if we had the histogram of a probability distribution and the area of the histogram was 1, then the areas of the rectangles measured probability. A Normal curve is such a probability distribution and by ensuring its total area is 1 then the areas under the Normal curve become probabilities. This means that we can interpret the result in the previous paragraph as follows:

(a) the probability of obtaining between 460 and 500 'heads' is 0.4938;

(b) the probability of obtaining between 460 and 520 'heads' is 0.8970;

(c) the probability of obtaining between 510 and 520 'heads' is 0.1775.

10. Continuous variables on the Normal curve. So far our random variables have all been discrete – i.e. you can only have an exact number of 'heads'. Often, however, the variable can be continuous. Consider the heights of, say, 1000 men. Height is a continuous variable but if you draw the histogram of heights using perhaps classes of one centimetre then, if the area of the whole histogram is regarded as '1', the graph would be a probability distribution graph.

To see this, assume that 89 men fell in the class 180 – 181cm. Since areas are proportional to frequency then the area of this particular rectangle will be 0.089 of the total area under the curve. But if you were to select at random any one man, the probability he had a height between 180 and 181cm. would also be 0.089. In other words, the area of a histogram rectangle measures the probability that an item selected at random from the distribution will fall between the lower and upper rectangle class limits.

11. Frequencies equals areas equals probabilities. In the earlier part of the book we saw that in a properly constructed histogram frequencies and areas are essentially interchangeable. Now it must be understood that probabilities, too, enter the relationship so that probabilities are essentially interchangeable with both areas and frequencies. This means, for instance, that if a class has a frequency of 20 out of 100 – i.e. 0.2 of the total – then the area of its histogram rectangle will be 0.2 of the total area of the histogram and the probability that an item selected at random will fall within the class limits has a probability also of 0.2.

This interchangeability between the three measures will prove very useful when we turn to the next chapter on sampling theory.

12. The Normal curve and sampling theory. In sampling theory a useful measure is the spread of the random variable values of a normal distribution that embraces the large majority of the items of the distribution. Looking at Appendix 3 it will be seen that the proportion of a normal distribution that lies within the -2σ and $+2\sigma$ range is $2 \times 0.4772 = 0.9544$. To all intents and purposes this is 95 per cent of the total distribution, or 19/20ths. So, if an item were selected at random out of the distribution there would be only one chance in 20 of its value lying outside this range.

This one-in-twenty chance is an arbitrary but useful level for expressing when an event is reasonably certain. In other words, if there is only one chance in twenty of the value of an item taken at random lying outside the range, then it is reasonably certain that the value of an item taken at random will lie inside the range.

If a higher degree of certainty is required then the range -3σ to $+3\sigma$ can be taken when there is a probability of $2 \times 0.4987 = 0.9974 -$ i.e. 399 in 400 $-$ of the value of an item selected at random lying inside this range. Conversely there is a one-in-four-hundred chance that the value will lie *outside* this range.

Progress test 21

(*Answers in Appendix 4*)

1. Assuming that the weights of 10,000 items are normally distributed and that the distribution has a mean of 115kg and a standard deviation of 3kg:

(*a*) how many items have weights between:
  (*i*) 115 and 118kg;
  (*ii*) 112 and 115kg;
  (*iii*) 109 and 121kg
  (*iv*) 106 and 124kg?

(*b*) if you had to pick one item at random from the whole 10,000 items, how confident would you be in predicting that its value would lie beween 109 and 121kg?

2. Steel rods are manufactured to a specification of 20cm length and are acceptable only if they are within the limits of 19.9cm and 20.1cm. If the lengths are normally distributed, with mean 20.02cm and standard deviation 0.05cm, find the percentage of rods which will be rejected as (i) undersize and (ii) oversize.   (*CIMA*)

3. Your company requires a certain type of inelastic rope which is available from only two suppliers. Supplier A's ropes have a mean breaking strength of 1,000kg with a standard deviation of 100kg. Supplier B's ropes have a mean breaking strength of 900kg with a standard deviation of 50kg. The distribution of the breaking strengths of each type of rope is normal. Your company requires that the breaking strength of a rope be not less than 750kg.

All other things being equal, which rope should you buy, and why?
   (*CIMA*)

4. Manufactured items are sold in boxes which are stated to contain a weight of at least 40 ounces. The actual weight in a box varies, being approximately normally distributed with mean 41.2 ounces and standard deviation 0.8 ounces.

Required:

(*a*) Calculate the proportion of boxes whose weight is between 40 ounces and 42 ounces.

(*b*) Calculate the weight below which 20 per cent of boxes fall.

(*c*) All boxes containing less than 40 ounces are scrapped at a cost of £1 per box. Calculate the scrapping cost associated with the sale of 100 boxes.

(*d*) To what mean weight should the box contents be adjusted, the standard deviation remaining unchanged, if only 1 per cent of boxes are to be scrapped?   (*ACCA*)

5. The independent probability of a passenger arriving for a booked flight on a Maxi service is 0.8.

The airline books 225 passengers and there are 195 seats available on a Maxi.

Use a Normal distribution approximation to find the probability that for a given flight more booked passengers arrive than there are seats available.

(Note that this is the second part of question 3, Progress test 20.)
   (*CIMA*)

22
Sampling theory

With the knowledge we now have of probability distributions and the connection between frequencies, areas and probability, we are in a position to look at sampling theory.

Sampling distributions

Prior to looking at probability distributions our statistical studies concentrated on maximising our comprehension of the groups of figures we collected. Now we go a step further and maximise our comprehension, not just of the figures we collect, but of the *populations we collect them from*. Indeed, this is more a leap than a step, for consider what it implies. What we have been doing so far has been to take, say, a collection of horses in a field and find ways of describing the horses we see. Now we shall be looking at the horses in a field and, on the basis of that sample, describing *all* horses, including the thousands we cannot see.

1. Statistical inference. Now, drawing a conclusion about a population from a sample is called *statistical inference*, and on the face of it an attempt to infer statistical facts about a whole population from a sample comprising only a small proportion of that population can be no more than rash guesswork. This, however, is not so for, although there will always be some degree of uncertainty, where large numbers are concerned the laws of probability are surprisingly consistent. Naturally, a freak result can always be thrown up by chance. But in practice such a highly improbable event can be ignored (though not forgotten), just as we can ignore the small but irremovable probability of being knocked down in the street when we arranged to meet someone for lunch.

2. Basic points. Before starting on our new theory a few important points should be noted:

(a) A sample, as we saw in 3:**2**, is a group of items taken from a population for examination. For sampling theory to work it is essential that *all items in a sample are selected at random.* Failure on this point means the laws of probability will not underlie our technique and the confidence they would otherwise bestow on our conclusions will not be present.

(b) Since, as was said above, we can never be absolutely certain of our conclusions about the population at large, then every time we draw such a conclusion we must *indicate the probability of our conclusion being in error.*

(c) The sampling theory that relates to small samples is slightly different from that which relates to large samples. In this book we are concerned only with the latter and so all the theory in this chapter will relate to *samples of 40 or above.*

(d) In sampling theory discussion two terms are sometimes used. These are:

(i) *parameter* – a parameter is any measure relating to a population;

(ii) *statistic* – a statistic is any measure relating to a sample.

(e) In sampling work the following symbols are used:

(i) $\bar{x}$ for the sample mean and s for the sample standard deviation;

(ii) μ for the population mean and σ for the population standard deviation.

3. Sample and sampling distributions. At this point it is vital that one thing is clearly understood and that is the difference between a sample distribution and what we shall call a sampling distribution:

(a) *Sample distribution.* So far we have confined ourselves to detailing carefully all the facts about a number of items, these items often being in effect a sample of all the items we could have selected from a larger population. When, in fact, we are dealing with the distribution of a sample then, not surprisingly, we refer to it as a *sample distribution*.

(*b*) *Sampling distribution.* Now, however, imagine we take thousands of samples (thousands of samples, note, not a sample of thousands – the difference is vital) and select one piece of information from each sample. This can be anything you like – the heaviest item in the sample, the sample mean, the sample median, the lowest value. If you then treat all these single figures, one from each sample, as a distribution in its own right then you obtain a sampling distribution. A *sampling distribution*, then, is the *distribution of values from a mass of samples, one value per sample.*

4. Distribution of a sample. Now let us assume that a population exists having the distribution depicted by the upper curve in Fig. 22.1, this population being 10,000 in number and having a mean of 20. If we were to take a random sample of 100, i.e. 1 per cent, from this distribution, what sort of sample distribution would we obtain?

For the sake of illustration let us say that 2,000 of the population fell in the class 15 to 20. In such circumstances we could expect to find about 1 per cent of 2,000, i.e. 20, of the items in our sample falling between these limits. Similarly, if another class contained 200 of the population we could expect our sample to contain 2 such items. In other words, we could expect our sample distribution to mirror the population distribution fairly exactly and so have a frequency curve similar to the lower one shown in Fig. 22.1. This, then, would be the *distribution of the sample*.

Next the reader should appreciate that if all the frequencies in a larger distribution are scaled down so as to give an identical distribution with smaller frequencies then the *means and the standard deviations of the two distributions will also be identical.* (To convince himself the reader should take any distribution he wishes, divide all the frequencies by an arbitrary constant, compute the means and standard deviations of the two distributions and then compare his results.) From this it follows that if a sample distribution exactly mirrors the population distribution then the *population* mean and standard deviation can be estimated directly from the *sample* mean and standard deviation.

5. Point estimate. When we want to give a single figure as an

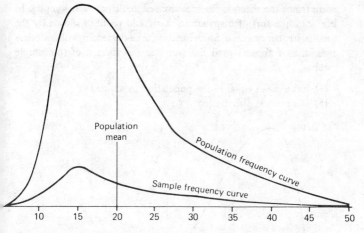

Figure 22.1 *Distributions of population and random sample taken from that population.* Note that the sample mean and dispersion will be approximately the same as the mean and dispersion of the population.

estimate of a population measure we say we are making a *point estimate*. The point estimate, then, of a population mean is given by the sample mean. However, since it is very rare for a sample distribution to mirror *exactly*, the population distribution, it is unlikely that such a point estimate would be exactly right. The big question is, then, just what potential margin of error can we expect when we make a point estimate of a population mean from a sample mean?

6. Sampling distribution of the sample mean. To answer this question assume that we take thousands of samples, each the same size, e.g. 100, and for each sample we compute the sample mean. These means will not, of course, all be identical because chance factors will result in some samples containing a slightly disproportionate number of larger values and some a slightly disproportionate number of smaller values. But for all that all the thousands of means will cluster around the population mean. Moreover, *these means will themselves form a quite separate distribution,* and this distribution is called the *sampling distribution of the sample* *mean* (since the mean is the measure selected from each sample). In Fig. 22.2 the sort of distribution that could result is shown by the smaller frequency curve. Studying this curve the reader will doubtless agree that it appears as if the sampling distribution of the sample mean:

(*a*) has a mean equal to the population mean; and
(*b*) is a normal distribution.

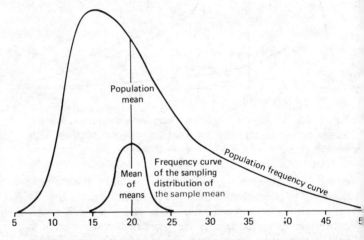

Figure 22.2 *Distribution of population and sampling distribution of the sample mean of samples taken from that population*

7. Symmetry of the distribution of the sample mean. Now the fact that this sampling distribution has a mean equal to the population mean can generally be accepted by readers. After all, if the means of all the samples taken from the same population were averaged it seems very probable that this overall average would indeed equal the population mean. And it would. But readers are sometimes sceptical that the distribution of the sample mean of a very skewed population is symmetrical – as it must be if it is to be a normal distribution. Yet it is symmetrical (provided, that is, the sample size is not less than around 40).

To try and gain some understanding of this important point consider a population in which 80 per cent of the items have a zero value and 20 per cent have a value of 1,000. Clearly, this rather unusual distribution is very skewed.

Now the mean of this population will, of course, be $(0.8n \times 0 + 0.2n \times 1,000)/n = 200$, and if a sample of 40 items contains 20 per cent (8 items) having a value of 1,000 then the sample mean would be 200. Imagine, however, two samples, one of which contains four items of 1,000 and another contains 12 items of 1,000. These two samples would have means of 100 and 300 respectively, i.e. they would be ranged the same distance either side of the population mean of 200. And the probabilities of these two samples is obtained from the binomial distribution formula, i.e. (where $x =$ number of items of 1,000):

$$P(x=4) = \binom{40}{4} \times 0.2^4 \times 0.8^{40-4} = 0.047$$

$$P(x=12) = \binom{40}{12} \times 0.2^{12} \times 0.8^{40-12} = 0.044$$

As can be seen, despite the skewness of the population the probability of obtaining a sample with a mean of 100 below the population mean is much the same as the probability of obtaining a sample with a mean 100 above. So although further calculations would show that the sampling distribution is indeed skewed, clearly, even in this extreme case, the skewness can be nothing like as pronounced as one might intuitively expect.

8. Sampling distribution of the sample mean is normal. Readers will unfortunately have to accept that our sampling distribution is, in fact, normal as well as symmetrical. However, their past work with probability theory will doubtless reassure them that sample means which deviate from a population mean by larger and larger values will become increasingly rare.

9. Standard deviation of the sampling distribution of the sampling mean. Given that it is now accepted that the sampling distribution of the sample mean is normal, the next question is, what is its standard deviation?

Now, theoretically this figure could be found in the usual way, i.e. by taking the sample mean as the random variable and using the ordinary formulae for finding the standard deviation. But if this were done in practice the work involved would be so extensive as not to be justified. Fortunately, no such work is necessary, for there is a connection between the standard deviation of the sampling distribution of the sample mean, the standard deviation of the population from which the samples are drawn, and the sample size. This connection is such that:

$$\text{Standard deviation of the sampling distribution of the sample mean} = \frac{\text{Standard deviation of the population}}{\sqrt{\text{Sample size}}}$$

At first glance this does not seem to improve matters, as we still need to know the standard deviation of the population. However, we saw earlier that the standard deviation of a population is approximately equal to the standard deviation of a sample taken from that population. The formula can therefore be rewritten as:

$$\text{Standard deviation of the sampling distribution of the sample mean} = \frac{\text{Standard deviation of the sample}}{\sqrt{\text{Sample size}}}$$

This means that we need to take only *one sample* in order to find the standard deviation of the whole of the sampling distribution of the sample mean of a whole host of samples.

10. Standard error of the sample mean. To save writing 'standard deviation of the sampling distribution of the sample mean' every time we will call this value the *standard error of the sample mean* and symbolise it as $\sigma_{\bar{x}}$. So our formula can be written as:

$$\sigma_{\bar{x}} = \frac{s}{\sqrt{n}}$$

This shorthand convention of using the term 'standard error' to stand for 'standard deviation of the sampling distribution' is so convenient that it is used in other contexts involving sampling distributions – though in every case we must say what sampling distribution we are talking about. Readers sometimes think that a standard error is something rather special, but just as a colonel is no more than a particular kind of army officer, so a standard error is no more than a particular kind of standard deviation. So as well as the standard error of the sample mean we can have the standard error of the median, and later we will have the standard error of the proportion. Its claim to a special name arises simply because it is a standard deviation that provides a measure of the potential error involved in estimating a population figure from a sample figure.

Estimating population means

Having found the standard error of the sample mean of a given sampling distribution, how can we use it?

11. The standard error and the population mean. We know from the previous chapter that 95 per cent of the items in a normal distribution lie within two standard deviations of the mean of the distribution. And we know from the previous section that a sampling distribution of the sample mean is a normal distribution with a mean equal to the population mean and a standard deviation we call the standard error. It follows, then, that 95 per cent of the means of *all* samples must lie within two standard errors of the population mean. And from this in turn we can say that if we take a single sample, then 19 times out of 20 the sample mean will lie within two standard errors of the population mean. From this it follows that *19 times out of 20 the population mean cannot be more than two standard errors away from the sample mean.*

12. Estimating the population mean. Let us look again at what we are saying from the beginning. We take a sample from a large population. We find the mean of the sample and know it must be close to the population mean – so close in fact that we can say it is approximately the population mean.

Unfortunately, we do not know how great our error is in taking the sample mean as the population mean. However, if we compute the standard error from the standard deviation of our sample, we shall be able to say that 19 times out of 20 the population mean will be within two standard errors of the mean of the sample, i.e. the population mean lies within the sample mean $\pm 2\sigma_{\bar{x}}$.

Example

Assume that our distances in Table 7D were but a random sample taken from a much larger population of distances, and that we wish to estimate the true mean distance of that population.

The mean distance of our sample, Table 7D, was 454½ kilometres and its standard deviation was 27 kilometres.

∴ the best estimate of the mean and standard deviation of the population is also 454½ and 27 kilometres.

$$\therefore \text{ standard error } (\sigma_{\bar{x}}) = \frac{s}{\sqrt{n}} = \frac{27}{\sqrt{120}} = 2.46 \text{ kilometres}$$

Now, 19 times out of 20 the population mean is within two standard errors of the mean of the sample.

∴ 19 times out of 20 we can say the population mean is between 454½ ± 2 × 2.46 ≈ 449.6 and 459.4 kilometres.

Note that here we are not giving a point estimate but rather a range estimate.

13. Confidence levels. It has been frequently emphasised that our conclusion will be correct 19 times out of 20. This of course is because 95 per cent (nineteen-twentieths) of all the sample means fall within two standard errors of the population mean. If the chance of being wrong 1 time out of 20 is too great a risk to take, we can be safer still by widening the range. If we extend it to three standard errors, then since 99¾ per cent of all the sample means will fall within three standard errors of the population mean we can be sure of being correct 399 times in 400 (99¾ per cent).

These different levels of certainty are known as *confidence levels*. Any range estimate of a population mean must always indicate what level of confidence has been adopted.

14. Confidence limits and interval. In **13** above we saw how changes of confidence level led to changes in the limits of the range within which the population mean can lie. The limits are called the *confidence limits* since they are the limits determined by the chosen level of confidence. Logically, too, the interval between these limits is called the *confidence interval*. In the example given above, the 95 per cent confidence limits are 449.6 and 459.4 and the 95 per cent confidence interval is 459.4 − 449.6 = 9.8 kilometres.

Note that the higher the confidence level, the greater the confidence interval.

15. Summary: estimating a population mean.

(a) Take a random sample of n items.

(b) Compute the sample mean ($\overline{x}$) and the sample standard deviation s.

(c) Compute the standard error of the sample mean ($\sigma_{\overline{x}}$) from the formula $\sigma_{\overline{x}} = s/\sqrt{n}$.

(d) Choose a confidence level, e.g. 95 per cent.

(e) Find from Normal curve tables the z value for this confidence level (e.g. 2).

(f) Estimate the population mean as $\overline{x} \pm z \times \sigma_{\overline{x}}$ (e.g. $\mu = \overline{x} \pm 2\sigma_{\overline{x}}$).

16. Effect of population size.
It should be noted that nothing has been said at any time about the size of the population in this theory. This means that the *accuracy of our estimate is quite independent of population size.*

In other words, contrary to what seems common sense, a population of 1,000,000 calls for no bigger sample than a population of 10,000. Our accuracy depends solely on sample size and the variability of the characteristic measured.

Note: Strictly speaking this is only true if the sample size is an insignificant proportion of the population size. If is is not an insignificant proportion, the real accuracy is actually *greater* than that claimed by the theory.

17. Combined populations.
Finally in this section it should be noted that if two independent populations equal in size are combined, then the combined mean and variance are given by:

(a) combined mean = sum of individual means;

(b) combined variance = sum of individual variances.

Example

An exercise in which journey times were taken at random gave the following results:

	Mean	Standard deviation
By road to the station	20 mins	8 mins
By rail to work	40 mins	6 mins

What is the overall average journey time? If journey times are normally distributed, can a traveller be at least 95 per cent confident his journey will not exceed 80 minutes?

Solution

Mean overall journey time $= 20 + 40 = 60$ minutes
Overall variance $= 8^2 + 6^2 = 100$ minutes
∴ Standard deviation $= \sqrt{100} = 10$ minutes

Now since the journey time will *exceed* 'mean $+ 2\sigma$' on only $2\frac{1}{2}$ per cent of all occasions, one can be $97\frac{1}{2}$ per cent confident that the journey time will not exceed $60 + (2 \times 10) = 80$ minutes. So a traveller can certainly be more than 95 per cent confident that his journey time will not exceed 80 minutes.

Estimating population proportions

There are occasions in statistics when information cannot be given as a measure, e.g. kilometres, tonnes, minutes, pence, examination marks, but only as a *proportion*, such as males in a group of people, left-wing voters in an electorate, or defective production in total production. In these cases we are faced with estimating the *population proportion* from a single sample.

18. Sampling distribution of the sample proportion. If we again took a large number of samples of (say) 100 people from a particular population, we should not always find that we had the same number of males in every sample. If the population contained slightly more females than males, we might find that our samples contained anything from 30 to 60 males, i.e. proportions ranging between 0.3 and 0.6.

Assume that we now construct the histogram of the distribution of the proportions from this very large number of samples. We will, of course, again be depicting a sampling distribution, this time the *sampling distribution of the sample proportion*. In Fig. 22.3 such a sampling distribution is shown. As can be seen, the distribution is

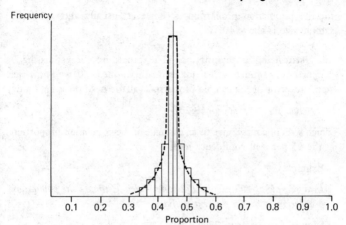

Figure 22.3 *Sampling distribution of a proportion.* This graph assumes the population proportion to be 0.45. It shows the frequency curve derived from the histogram of the various proportions found in a large number of samples.

again normal and has a mean proportion equal to the population proportion.

19. Standard error of the sample proportion. Now, we can estimate the population proportion from our sample by using the sample proportion, but unless the standard deviation of this sampling distribution – the *standard error of the sample proportion* – can be found, there is no way of determining the possible error in our estimate. Our other standard error formula is of no value here, since it is not possible to determine the standard deviation of the sample proportion from a single sample. How, for instance, could you calculate the standard deviation of a sample where 43 out of 100 people were male? Fortunately, there is yet another formula available which states that:

Standard error of the sample proportion $(\sigma_p) = \sqrt{\left(\dfrac{pq}{n}\right)}$

where p = population proportion, $q = 1 - p$, and n = sample size.

Unfortunately this formula calls for the population proportion – the very thing we are trying to find! Substitution of the *sample* proportion for the population proportion is, however, usually allowed if the sample size is above 40.

20. Estimating a proportion. The procedure for estimating a proportion is similar to that for estimating a mean. If we know the standard error of the sample proportion, all we need do is compute:

Actual sample proportion $\pm 2\sigma_p$

which will, of course, give us an estimate of the population proportion at the 95 per cent confidence level.

Example

In a sample of 400 people, 172 were males. Estimate the population proportion at the 95 per cent confidence level.

Sample proportion $= 172/400 = 0.43$

$$\sigma_p = \sqrt{\left(\frac{pq}{n}\right)}$$

$$= \sqrt{\left(\frac{0.43 \times (1 - 0.43)}{400}\right)}$$

$$= \sqrt{\left(\frac{0.43 \times 0.57}{400}\right)}$$

$$= 0.0248$$

∴ Estimate of population proportion at the 95 per cent confidence level $=$ Actual sample proportion $\pm 2\sigma_p$

$$= 0.43 \pm 2 \times 0.0248$$
$$= 0.3804 \text{ and } 0.4796$$
$$\simeq \text{between 38 per cent and 48 per cent}$$

21. Summary: estimating a population proportion.

(*a*) Take a random sample of n items, n being 40 or more.
(*b*) Compute the sample proportion.
(*c*) Compute the standard error of a proportion (σ_p) from formula $\sigma_p = \sqrt{(pq/n)}$, using the sample proportion for p.
(*d*) Choose a confidence level, e.g. 95 per cent.
(*e*) Find from Normal curve tables the z value for this confidence level (e.g. 2).

(f) Estimate the population proportion as *actual sample proportion* $\pm z \times \sigma_p$ (e.g. sample proportion $\pm 2\sigma_p$).

Comparing this procedure with that for the estimation of the population mean (**15**) will show that the two are virtually identical.

Testing a hypothesis

It frequently happens in statistical work that some factor is asserted to be true, yet when a random sample is taken it turns out the sample data does not wholly support the assertion. The difference could be due to either:

(a) the original belief being wrong; or
(b) the sample being slightly one-sided – as virtually all samples are to some degree.

Clearly tests are needed to distinguish which is the more likely possibility. In this section we see how we can test if assertions regarding population means and proportions are likely to be wrong or not.

22. Statistical significance. Tests of this nature are based on ascertaining whether or not any difference between two figures can be explained as being due to reasonable chance factors. Where a *difference cannot be explained as being due to chance* – i.e. some other factor is needed to explain the difference – we say the *difference is statistically significant.* Note that this use of the word 'significant' differs from the day-to-day use of the word where it usually means 'important'. Thus, we would say that there was a significant difference between winning £1,000 and losing £1,000 at roulette. However, such a difference is not statistically significant since it can easily be explained by chance (indeed, it is). Conversely, if the average weight of city dwellers were 2 ounces greater than country dwellers we would probably say the difference in practice was not significant whereas if statistical tests showed this could not be explained by chance alone it would be deemed to be statistically significant.

23. Significance tests. Any test to see if a difference is statistically significant or not is termed a *significance test.* This section, then, is concerned with significance tests. And to illustrate such tests we will look specifically at the case where a sample mean differs from an asserted population mean.

24. The null hypothesis. The approach to this test is to make the hypothesis, or assumption, that *there is no contradiction between the asserted figure and the sample figure*, i.e. between the asserted parameter and the sample statistic, and that the difference can, therefore, be ascribed solely to chance. This hypothesis is called the *null hypothesis*, and the object of the test is to see whether the null hypothesis can be rejected or not.

25. Testing the null hypothesis – theory. To do this in the case of our test relating to the population and sample mean we make use of the sampling distribution of the sample mean. If the population mean is in fact the figure asserted then 95 per cent of sample means will fall within two standard errors of this figure. So all we need to do, then, is to find the range of two standard errors either side the asserted population mean and see if the sample mean falls within this range. If it does then the difference between the asserted population mean and the sample mean can be wholly explained by chance. If it does not, then the difference *cannot* be wholly explained by chance – unless, that is, the sample mean happens to be one of the one-in-twenty means that will inevitably lie outside this two standard error range. And if the difference between the asserted mean and the sample mean cannot be explained by chance, then it is unlikely that the asserted mean is the correct population mean.

Note that in this latter case we say we 'reject the null hypothesis'. In other words, *rejecting a null hypothesis* means that we reject the assertion 'there is no contradiction between the asserted population mean and the sample mean', and conclude the contradiction is so great that one mean or the other is incorrect. Since the sample mean (if properly found) is an indisputable fact, it tends to win the argument.

26. Testing the null hypothesis – procedure. The basic procedure to test the null hypothesis is as follows (though *see* **37** for a fuller description of the procedure):

(a) Find the sample mean and standard deviation.

(b) State the null hypothesis.

(c) Use the sample standard deviation to estimate the po..... standard deviation (since this will be all you will have) and compute the standard error – i.e. $\sigma_{\bar{x}} = s/\sqrt{n}$

(d) Find the range within which the sample mean should fall – i.e. asserted mean ± two standard errors.

(e) Look to see if the sample mean does fall within the range or not.

(f) If it does, do *not* reject the null hypothesis. If it does not, reject the null hypothesis.

Example

Assume that Table 7D relates to the distance travelled by the 120 salesmen taken at random from a very much larger field force. Someone now asserts that the mean distance travelled by all the salesmen in the field force is 460km. Our sample mean, however, is 454½km. Can this assertion be maintained at the 95 per cent level of confidence?

In Table 7D $s = 27$km, $n = 120$.

(a) Sample mean equals 454½. Sample standard deviation equals 27.

(b) Null hypothesis: no contradiction that cannot be explained by chance between the asserted population mean of 460 and the sample mean of 454½.

(c) Standard error $= s/\sqrt{n} = 27/\sqrt{120} = 2.46$km.

(d) Range within which sample mean should fall equals $460 \pm 2 \times 2.46 = 455.08$ to 464.92.

(e) Sample mean of 454½ does *not* fall within this range.

(f) So the null hypothesis is rejected.

So we conclude that it is unlikely that the asserted mean of 460km is correct.

27. Assertion correct until proved wrong. It should be noted that when testing a hypothesis the asserted figure must be taken as correct until there is sufficient evidence to reject it. It is for this reason that we checked above if the sample mean fell within the two standard error range either side the asserted mean – not if the asserted mean fell within the range two standard errors either side of the sample mean.

In this particular case it is the same either way, but the principle of accepting the population figure until proved wrong should be observed right from the beginning.

28. Non-rejection of the null hypothesis. If the null hypothesis is rejected we conclude that the population mean is not the figure originally asserted. But if the null hypothesis is *not* rejected it is important to appreciate that we do *not* conclude the population mean *is* the figure asserted. Non-rejection of the null hypothesis only signifies that there is no evidence that the true mean is not as asserted; it does not mean there is evidence that it is correct. Thus if, in the above example, our sample mean had been 456, we would not have rejected the null hypothesis, although obviously the true population mean could as well have been, say, 455 or 458 – or virtually any figure in the 450s – as the asserted figure of 460. Testing a hypothesis cannot result in proof that the asserted figure is true. It can only show that such a figure is probably false.

29. Significance level. When we reject the null hypothesis there is, of course, always a chance that our sample actually *is* one of the more extreme samples from a population which really does have the asserted mean and so we will have wrongly rejected the hypothesis. It is important, then, to indicate just how confident we are that our rejection is valid. By using two standard errors we can be confident that 19 times out of 20 (i.e. 95 per cent) we will be correct since only once in 20 occasions will we be unlucky enough to have a sample that misleads us. As in the case of estimating the population mean (**13**) it is necessary for us to indicate the degree of confidence we have in our decision. In that previous instance we stated that by using two standard errors we were 95 per cent confident that our estimate was correct. In the case of significance tests we state how probable it is that we have made an error. Obviously, if 95 per cent of the time we are correct then 5 per cent of the time we will have made an error. To this probability that we are in error in rejecting the null hypothesis we give the name *significance level*. In significance testing, then, we use significance levels and in the current instance we would say that our decision to reject the null hypothesis is significant at the 5 per cent level.

Note that since a significance level is simply the converse of a confidence level we can find the former level by the formula:

Significance level = 100 − level of confidence

30. Confidence interval and limits. In step (*d*) of our procedure we calculated the range within which the sample mean should lie, given our chosen level of confidence. Again, as in **14**, this range is called the *confidence interval*, and its limits at each end the *confidence limits*.

31. Significance level and the risk of rejecting a true hypothesis. As can be seen, our level of confidence indicates that the population mean could not be 460km, unless our sample were the exceptional one in twenty. But what if the sample *were* the odd one in twenty? In that case we would have rejected the null hypothesis when it might be true and when 460km could, in fact, have been the true population mean.

The significance level indicates the risk one takes of rejecting a null hypothesis that might well be correct. To dismiss the assertion that 460km was the population mean might lead to action being taken that could result in serious difficulties should the population mean turn out to be 460 after all. Under such circumstances it may be considered that the risk of being wrong one in twenty times is too great and that only a risk of one in four hundred, at most, is justifiable. In that case a significance level of $\frac{1}{4}$ per cent would be selected, i.e. the standard error would be multiplied by 3 when computing the confidence interval.

32. Significance level and the risk of not rejecting an incorrect hypothesis. However, lowering the significance level can result in the opposite error, since it means that one risks clinging to an asserted mean of 460 when in fact it is incorrect. In the example above, use of the $\frac{1}{4}$ per cent significance level brings the lower confidence limit down to $460 - 3 \times 2.46 = 452.62$. Since our sample mean is above this ($454\frac{1}{2}$) we would not reject the null hypothesis.

33. Type I and Type II errors. Clearly, we are in something of a dilemma when it comes to selecting a significance level. Whatever we do we risk an error one way or the other. These two complementary risks have been given the classifications of Type I and Type II errors and can be defined as follows:

(a) a *Type I error* is the *error of rejecting a hypothesis when it is in fact true*;

(b) a *Type II error* is the *error of not rejecting a hypothesis when it is in fact false*.

Of course, we can only decrease the risk of one type of error by increasing the risk of the other. In practice, therefore, before starting a significance test we must decide on the importance of each type of error and then set the significance level accordingly. For example, an assertion that a given maximum absorption of radioactivity was safe would be tested at a very low level of significance since refusing to accept the assertion even if it were true would be far less disastrous than accepting it should it prove that the absorption figure actually was not safe.

In short, the selection of the significance level depends on which error is considered the graver: to reject a hypothesis which may be true, or to fail to reject one that is wrong.

Finally, note that it is essential to select the significance level *before* testing the hypothesis. Errors of interpretation could arise if one measured the deviation of the sample mean from the asserted population mean first, and then computed the chance of such a difference arising.

34. One-tail and two-tail probabilities. At this juncture a rather subtle point arises. In significance testing one should be quite sure just what is being tested. Can you, for instance, see the difference between the two following assertions

(a) The average weight of each box of my sausages is 35kg.

(b) The average weight of each box of my sausages is not less than 35kg.

Clearly, in (a) what is being asserted is that the average is 35kg – neither more nor less – while what is being asserted in (b) is that the average weight is not less than 35kg, *though it could be more.* Now if the average weight *were* 35kg, the standard deviation were, say, 2kg and samples having sizes of 64 boxes were taken, then a distribution of the sample mean would be obtained having a mean of 35kg and a standard error of $2/\sqrt{64} = 0.25$kg. So 95 per cent of all samples will have means that lie inside the range $35 \pm 2 \times 0.25 = 34.5$ to 35.5kg – and 5 per cent will lie *outside* this range. So there will be only 1 chance in 20 of a sample having a mean weight below 34.5kg or above 35.5kg. But note, *this 5 per cent is made up of that area under the combined two tails of a normal curve that lie below 34.5kg and above 35.5kg (see* Fig. 22.4). However, if we only want the probability of taking a sample the mean of which lies *below* 34.5kg then we must only measure the area below the lower *one tail* of the curve, and this obviously, is half 5 per cent, i.e. $2\frac{1}{2}$ per cent.

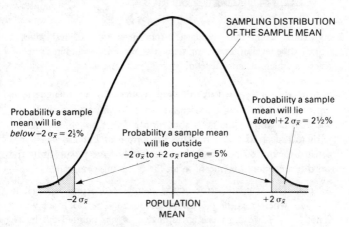

Figure 22.4 *One-tail and two-tail probabilities*

35. One- and two-tail tests. Now let us apply this to significance tests. If assertion (*a*) in the previous paragraph is to be tested, then we will be equally interested in a population mean that is above 35kg as one that is below. If our confidence interval is 34.5 to 35.5kg, we would be testing the assertion at the 5 per cent level of significance since 5 per cent of sample means would lie outside this range. If, however, assertion (*b*) is to be tested, we would only be interested in a population mean that fell *below* 35kg, and if our confidence limit were 34.5kg, then we would be testing the assertion *at the $2\frac{1}{2}$ per cent level of significance* since only $2\frac{1}{2}$ per cent of sample means would lie below this limit (*see* Fig. 22.4 again). For obvious reasons an assertion (*a*) kind of test is called a *two-tail test* and an assertion (*b*) kind of test is called a *one-tail test*.

What all this means is that when deciding on a significance test it is necessary to decide just what it is you want to test; whether the true mean is different from the asserted mean (which calls for a two-tail test) or whether the true mean lies specifically at one side of the asserted mean (which calls for a one-tail test). In the latter case, note that for a given standard error, z, the level of significance of a one-tail test is

$$\frac{\text{Level of significance of a two-tail test}}{2}$$

e.g. since using two standard errors in a two-tail test gives a significance level of 5 per cent, using two standard errors in a one-tail test gives a significance level of $5/2 = 2\frac{1}{2}$ per cent.

36. Testing a hypothesis involving proportions. So far we have only considered testing a hypothesis about a population mean. If now we wish to test a hypothesis about a *population proportion*, e.g. we wish to test an assertion that the proportion of Labour voters in a constituency is 60 per cent, then we simply have to substitute the word 'proportion' for 'mean' in all the foregoing paragraphs.

Note: The standard error of the sample proportion is equal to $\sqrt{pq/n}$ where p is the population proportion (*see* **19**). Since we do not know p we must use an estimate. As was pointed out in **27**, when testing a hypothesis the asserted parameter must be used – i.e. p must be estimated as the *asserted* population proportion, and not as the sample proportion.

37. Summary: testing a hypothesis. When testing a hypothesis the object is to test whether the difference between an asserted mean (proportion) and a sample mean (proportion) is significant or not. The procedure is:

(*a*) Compute any necessary statistics (e.g. standard deviation of the sample) and find the relevant standard error ($s/\sqrt{n}$ or $\sqrt{pq/n}$)

(*b*) State the null hypothesis.

(*c*) Decide on:

(*i*) the relative balance between making a Type I error and making a Type II error, and accordingly select the significance level;

(*ii*) whether a two-tail or a one-tail test is called for.

(*d*) Look up in Normal curve tables the value of z for the selected confidence level and type of test

(*e*) Compute the relevant confidence limits – i.e. asserted mean $\pm z \times \sigma_{\bar{x}}$ or asserted proportion $\pm z \times \sigma_r$

(*f*) Check where the sample mean (proportion) falls:

(*i*) if *outside* these limits, reject the null hypothesis at the selected level of significance – i.e. the population mean (proportion) is not as asserted;

(*ii*) if *inside* these limits, do not reject the null hypothesis at the selected level of significance – i.e. there is no evidence to disprove the asserted mean (proportion) and it could be as asserted.

Testing differences between means and proportions

It is quite common in statistical work to be confronted with two distinct populations which seem likely to have virtually identical means. For example, cats in the north of England are probably the same height as those in the south, i.e. the mean heights of cats in the two populations are the same. Nevertheless, if a sample were taken from each population it would be unlikely that the two sample means would be identical. How, then, could we tell whether the difference between the sample means was due solely to chance factors, or to a real difference between the two population means? In other words, how can we tell whether the difference between the means is significant?

38. Sampling distribution of the difference between means.
Assume that the two population means were the same. In that case, if a great many pairs of samples were taken and the difference found between the means of each pair (always deducting the mean heights of the southern cats from the mean heights of the northern cats), the differences would be found to be small – indeed, there would be occasions when there was no difference. When there were differences,

about half of them would be plus and half minus. On only a very few occasions would there be large differences, plus or minus.

If these differences were graphed it would be found that once again the sampling distribution would follow a normal curve, this time one with a mean of zero and extending over plus and minus values (*see* Fig. 22.5). And, again, 95 per cent of the differences would lie within two standard errors of the mean of zero; in other words, *95 per cent of the differences would not exceed two standard errors.*

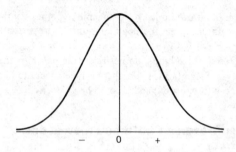

Figure 22.5 *Sampling distribution of the difference between means*

39. Testing a difference between means. The original problem now becomes one of testing to see whether or not an actual difference found exceeds two standard errors in a distribution of differences between means. If it does not exceed two standard errors, the difference can be set down to chance and one can conclude there is no evidence to prove that the two populations do not have the same mean.

First, however, it is necessary to find the standard error in a distribution of differences between means. This figure is given by the formula:

$$\sigma_{(\bar{x}_1 - \bar{x}_2)} = \sqrt{(\sigma^2_{\bar{x}_1} + \sigma^2_{\bar{x}_2})}$$

where $\sigma_{(\bar{x}_1 - \bar{x}_2)}$ = the standard error of the difference between means.

We check, therefore, to see whether or not the actual difference exceeds $2\sigma_{(\bar{x}_1 - \bar{x}_2)}$. If it does, the population means are most unlikely to be the same. In statistical jargon we should say that the difference between the means is significant at the 5 per cent level of significance, and therefore the populations do not have the same mean. Of course, other significance levels could be selected, in which case $\sigma_{(\bar{x}_1 - \bar{x}_2)}$ would need to be multiplied by the appropriate value which as usual is found from Normal Curve tables.

Example

Assume that:

(1) *North of England cats* (x_1): sample mean 25cm; standard deviation of sample $6\frac{1}{4}$cm; sample size 100;

(2) *South of England cats* (x_2): sample mean 24cm; standard deviation of sample 6cm; sample size 144.

Now $\sigma_{\bar{x}} = \dfrac{s}{\sqrt{n}}$

$$\therefore \ \sigma_{\bar{x}_1} = \frac{6.25}{\sqrt{100}} = 0.625\text{cm}$$

and $\qquad \sigma_{\bar{x}_2} = \dfrac{6}{\sqrt{144}} \quad = 0.5\text{cm}$

And using formula above:

$$\sigma_{\bar{x}_1 - \bar{x}_2} = \sqrt{(\sigma^2_{\bar{x}_1} + \sigma^2_{\bar{x}_2})} = \sqrt{(0.625^2 + 0.5^2)} = 0.8\text{cm}$$

Now, if the 5 per cent level of significance is required (and note that this is a two-tail test), limits are

$$2\sigma_{(\bar{x}_1 - \bar{x}_2)} = 2 \times 0.8 = 1.6\text{cm}$$

And the actual difference between means $= 25 - 24 = 1$cm. Since this 1cm is within the 1.6cm limit, the difference could have arisen through chance factors and is not significant at the 5 per cent level of significance. There is no evidence, therefore, that the cats in the two parts of England have different heights.

40. One-tail and two-tail tests. Again note we should really first consider if we are to use a one-tail or two-tail test. A glance at Fig. 22.4 will indicate that whereas 5 per cent of the differences will lie *outside* the two standard errors from the mean range if there are no population mean differences, only $2\frac{1}{2}$ per cent will lie *beyond* a given two standard error limit. In other words, in our example above we tested at the 5 per cent level of significance whether or not there was a significant difference between the heights of cats in the north of England and the height of cats in the south. If we had been testing whether northern cats were *taller* than southern cats, then a one-tail test would be involved and it would follow from our results that we could say that the difference between the means was not significant at the $2\frac{1}{2}$ per cent level of significance. (In this particular case, of course, it is obvious that if there is no significance at the 5 per cent level of significance there certainly cannot be any at a lower level but the principle still applies.)

41. Summary: testing a difference between means. The following is a summary of the procedure to test the difference between two means, i.e. to test whether or not the difference between two sample means is significant.

(*a*) Select the level of significance required on the usual basis.

(*b*) Find the standard error of the sample mean ($\sigma_{\bar{x}}$) for both samples.

(*c*) Compute the standard error of the difference between means by formula:

$$\sigma_{(\bar{x}_1 - \bar{x}_2)} = \sqrt{(\sigma^2_{\bar{x}_1} + \sigma^2_{\bar{x}_2})}$$

Note: Since $\sigma_{\bar{x}} = s/\sqrt{n}$, this standard error can be calculated in one step by using the formula:

$$\sigma_{(\bar{x}_1 - \bar{x}_2)} = \sqrt{\frac{s_1^2}{n_1} + \frac{s_2^2}{n_2}}$$

(*d*) Multiply $\sigma_{(\bar{x}_1 - \bar{x}_2)}$ by the appropriate z value for the level of confidence selected.

(*e*) Find the actual difference between the two sample means. If the difference is below the limit found in step (*d*), it is not significant. If it is above that limit, the difference is significant and the conclusion can be drawn (at the chosen level of significance) that the two populations have different means.

42. Testing a difference between proportions. Problems sometimes involve testing a difference between *proportions* in samples instead of means. This gives little trouble, since the procedure outlined above can still be applied except that the word 'proportion' should be substituted for the word 'mean'. Even the formula for the standard error of the difference between proportions is the same, although of course the symbols change slightly, i.e.:

$$\sigma_{(p_1 - p_2)} = \sqrt{(\sigma^2_{prop_1} + \sigma^2_{prop_2})}$$

Note: Since $\sigma^2_p = (\sqrt{(pq/n)})^2 = pq/n$ then the formula can be written as:

$$\sigma_{(p_1 - p_2)} = \sqrt{\left(\frac{p_1 q_1}{n_1} + \frac{p_2 q_2}{n_2}\right)}$$

Progress test 22

(*Answers in Appendix 4*)

1. Estimate the population mean (*a*) at the 95 per cent and (*b*) at the 99.75 per cent confidence levels, where the sample data is:

   (*i*) mean, 950kg; *s*, 15kg; sample size, 25;
   (*ii*) mean, 1.82cm; *s*, 0.8cm; sample size, 100;
   (*iii*) mean, 1.82cm; *s*, 0.8cm; sample size, 10,000.

2. Estimate, at the 95 per cent level of confidence, the population proportion where sample data is:

   (*a*) 61 males out of 100 people.
   (*b*) 6,100 males out of 10,000 people.
   (*c*) 26 defectives out of 49 parts.

3. If a large box of bolts having a mean weight of 28g per bolt and a standard deviation of 2g is added to a large box of nuts having a mean weight of 8g per nut and a standard deviation of $\frac{3}{4}$g, estimate the weight of a combined nut and bolt.

4. The result of a sample survey of 100 flowers of a particular type showed that the estimated mean flower height was 15cm±2cm at the 95 per cent level of confidence. The investigator decides that he needs an estimate which is within $\frac{1}{2}$cm of the true population mean at this level of confidence. What must his sample size be?

5. If you wished to have a confidence level of 99 per cent in any survey, what would $\sigma_{\bar{x}}$ need to be multiplied by?

6. A child welfare officer asserts that the mean sleep of young babies is 14 hours a day. A random sample of 64 babies shows that their mean sleep was only 13 hours 20 minutes, with a standard deviation of 3 hours. Test the officer's assertion at the 5 per cent significance level.

7. An election candidate claims that 60 per cent of the voters support him. A random sample of 2,500 voters show that 1,410 support him. Test his claim at the 1 per cent significance level.

8. (a) Calculate the probability that, if 30 per cent of items in a batch are defective, a random sample of 100 items will contain 40 per cent or more defectives.

(b) It is claimed that a process produces not more than 30 per cent defective. A random sample of 100 items from the process was found to contain 42 defectives. Investigate the validity of the claim.

(ACCA)

9. A sample of 200 fish of a particular species taken at random from one end of a lake had a mean weight of 20kg and a standard deviation of 2kg. At the other end of the lake, a sample of 80 fish showed a mean weight of $20\frac{1}{2}$kg and a standard deviation of 2kg also. An expert on fish claimed that these fish swam all over the lake and the two samples were therefore taken, in effect, from the same population. Test this assertion at the 5 per cent significance level.

10. A health official claims that the citizens of city A are fitter than those of city B, and in evidence shows that 96 out of 200 citizens of city A, selected at random, passed a standard fitness test as against only 84 out of 200 citizens of B. Do you think that, at the 5 per cent significance level, he has proved his claim?

11. What is the relevance of the binomial distribution to sampling theory?

Assignment

Write down the numbers from 1 to, say, 500 and against each number write a value (or mark it with or without an asterisk) and find the mean of all the values (or the proportion of numbers which have an asterisk). Then take a number of random samples (i.e. choose a number at random and record its value or if it has an asterisk) of around 40, find the mean (or proportion) of the sample and estimate the population mean (or proportion) at the $66\frac{2}{3}$ per cent level of confidence. Repeat the sampling a number of times and see if your results confirm that two out of three times your estimate is correct.

Appendix 1
Basic mathematical concepts

This appendix explains two of the mathematical concepts necessary for understanding part of this book. These explanations have been written as an appendix so that the reader who understands the concepts does not have his reading of the development of the statistical theory interrupted by material with which he is already familiar.

The Σ notation

The first of these concepts relates to the Σ notation.

1. Σ. Assume we have to add together all the heights of a group of people (e.g. we wish to find the average height). We could designate each height as h and put a suffix to show we were talking about first, second, third, etc. height, i.e. we could write $h_1 + h_2 + h_3 + h_4 \ldots$ However, adding together a series of figures occurs so often in mathematics that a special symbol exists to indicate this operation. The symbol is Σ (pronounced 'sigma') and means 'add together'. If, then, we write 'Σh' this can be interpreted as 'add together all the hs'.

2. **Where to start and stop.** Useful as this symbol is as it stands, though, it is limited to situations where we add all the figures in the series together. However, we often wish to add only a part of the series – e.g. the heights of the first six people or, using our suffix convention, h_1, h_2, h_3, h_4, h_5, h_6. How can we adopt the Σ symbol to indicate where we start and where we end?

This is really no problem. If we generalise the suffix number and call it, say, i, and then write the first value i takes below the Σ symbol and the last value above, we have the required information. For example, writing $\displaystyle\sum_{i=1}^{i=6}$ means 'add together all the hs starting at $i=1$, that is h_1, and stopping at $i=6$, that is, h_6'. (Note the figures are inclusive, i.e. both h_1 and h_6 are included in the group to be added.)

If we need to add all the figures then we often simply write n (where n is the total number of figures) at the top, e.g. $\displaystyle\sum_{i=1}^{n}$.

The equation of the straight line

The second concept to explain relates to the equation:

$$y = a + bx$$

3. Describing a straight line on a graph. Draw a straight line on a graph – any graph (providing it has linear scales) and any straight line. Now ring up your friend down the road and describe the line so that he can accurately reproduce it on his own as yet unmarked graph paper.

Well, first, of course, you will have to tell him what figures to put on the two axes – the x-axis, the horizontal axis, and the y-axis, the vertical axis. But that done you then have to enable him to draw your line on his paper.

This, you will probably decide fairly quickly, is not too difficult. You will probably begin by telling him where the line starts, i.e. tell him how far up (or down) the y axis he should go to find where the line begins, i.e. the value of y when $x=0$. Let's call this value a. Next, you look carefully at your line and then tell him by how much it rises (or falls) if you move from the starting point to the point where $x=1$. Let us call this b. If your friend now joins the starting point to this new point with a ruler, and then carries on drawing the line he has formed he will end up with your line on his graph.

4. b = rise per unit. Note, now, the fact that where you have a straight line then, if it rises (or falls) by b when you are moving from

$x = 0$ to $x = 1$, it will rise (or fall) by another b when you move from $x = 1$ to $x = 2$. In fact, each time x increases by 1, the line rises (or falls) by b. So the total rise of the line from the place where $x = 0$ to the end of the line is $b \times x$, or bx, where x marks the end of the line as measured along the x-axis.

5. $y = a + bx$. From this it follows that all you need to give your friend to draw your line is a and b – i.e. with these two values the straight line you have on your paper is defined exactly – a marking the start of the line when $x = 0$ and b the rise per unit of x. And note that the value of y can be found for any value of x by the equation:

$$y = a + bx$$

This equation is known as *the equation of the straight line* since with a and b given their values it describes the line fully.

6. No start or finish. Of course, if your friend in **3** never stops drawing, his line will go on for ever and so you will have to tell him where to stop. However, in mathematical terms the line *is* deemed to go on for ever – both as x increases and as x goes into negative values. The equation of the line, then, presumes no limits either way.

Appendix 2
Statistical formulae

Essential formulae

Frequency distributions
Arithmetic mean $(\bar{x})$

$$\bar{x} = \frac{\Sigma x}{n}$$

Range

Range = Highest value − Lowest value

Quartile deviation

$$\text{Quartile deviation} = \frac{\text{Upper quartile} - \text{Lower quartile}}{2}$$

Mean deviation

$$\text{Mean deviation} = \frac{\Sigma(x - \bar{x})}{n} \text{ (sign of } (x - \bar{x}) \text{ to be ignored)}$$

Standard deviation

$$s = \sqrt{\left(\frac{\Sigma(x - \bar{x})^2}{n}\right)}$$

Alternative formula:

$$s = \sqrt{\left(\frac{\Sigma x^2}{n} - \bar{x}^2\right)}$$

Variance

$$\text{Variance} = s^2$$

Coefficient of variation

Coefficient of variation $= \dfrac{s}{x} \times 100$

Pearson coefficient of skewness (Sk)

$$Sk = \dfrac{3(\text{Mean} - \text{Median})}{s}$$

Correlation

Regression lines

Regression line of *y on x*

Line equation: $y = a + bx$

$$b = \dfrac{n \times \Sigma xy - \Sigma x \times \Sigma y}{n \times \Sigma x^2 - (\Sigma x)^2}$$

$$a = \dfrac{\Sigma y - b \times \Sigma x}{n}$$

where n = the number of pairs.

Regression line of *x on y*

Line equation: $x = a + by$

Interchange x and y in the above formulae (*see* 13:**5**)

Correlation (r)

$$r = \dfrac{n \times \Sigma xy - \Sigma x \times \Sigma y}{\sqrt{(n \times \Sigma x^2 - (\Sigma x)^2)(n \times \Sigma y^2 - (\Sigma y)^2)}}$$

Rank correlation

Spearman's coefficient of rank correlation $= 1 - \dfrac{6\Sigma d^2}{n(n^2 - 1)}$

where n = the number of pairs, and d = the difference between rankings of the same item in each series.

Standard errors

Standard error of the mean (σ_x)

$$\sigma_{\bar{x}} = \dfrac{s}{\sqrt{n}}$$

Standard error of a proportion (σ_p)

$$\sigma_p = \sqrt{\left(\frac{pq}{n}\right)}$$

where p = sample proportion, and $q = 1 - p$.

Standard error of the difference between means ($\sigma_{(\bar{x}_1 - \bar{x}_2)}$)

$$\sigma_{(\bar{x}_1 - \bar{x}_2)} = \sqrt{(\sigma^2{}_{\bar{x}_1} + \sigma^2{}_{\bar{x}_2})}$$

Standard error of the difference between proportions ($\sigma_{(p_1 - p_2)}$)

$$\sigma_{(p_1 - p_2)} = \sqrt{(\sigma^2{}_{p_1} + \sigma^2{}_{p_2})}$$

Index numbers

Weighted aggregative price index

$$\text{Index} = \frac{\Sigma(p_1 \times w)}{\Sigma(p_0 \times w)} \times 100$$

Laspeyre price index

$$\text{Index} = \frac{\Sigma(p_1 \times q_0)}{\Sigma(p_0 \times q_0)} \times 100$$

Paasche price index

$$\text{Index} = \frac{\Sigma(p_1 \times q_1)}{\Sigma(p_0 \times q_1)} \times 100$$

Useful formulae

Weighted average

$$\text{Weighted average} = \frac{\Sigma xw}{\Sigma w}$$

Geometric mean

$$\text{GM} = \sqrt[n]{(x_1 \times x_2 \times x_3 \times \ldots x_n)}$$

Alternative:

$$\text{Log GM} = \frac{\Sigma \log x}{n}$$

Harmonic mean

$$HM = \frac{n}{\sum \frac{1}{x}}$$

Bayes' Theorem

$$P(E|S) = \frac{P(E) \times P(S|E)}{\sum\limits_{i=1}^{n} (P(E_i) \times P(S|E_i))}$$

where S is the subsequent event and there are *n* prior events, E.

Standard normal distribution

$$z = \frac{x - \mu}{\sigma}$$

Price relative

$$\text{Price relative} = \frac{p_1}{p_0} \times 100$$

Weighted average of price relatives index

$$\text{Index} = \frac{\sum (\frac{p_1}{p_0} \times 100 \times w)}{\sum w}$$

Base changing

$$\text{New index number} = \frac{\text{Old index No.}}{\text{Old index No. of new base period}} \times 100$$

Asset revaluation

$$\text{New valuation} = \text{Original value} \times \frac{\text{New price index}}{\text{Original price index}}$$

Average annual change

$$\text{Average annual percentage change} = 100(\sqrt[n]{\text{later figure/earlier figure}} - 1)$$

Trend line

$$y = a + bd$$

where $y=$ the variable for which trend is required, $d=$ the deviation in time from the midpoint of the time series, $a=\bar{y}$, and $b=\Sigma yd/\Sigma d^2$.

Subdivision of the trend line

Sub-period value $y' = \dfrac{a}{n} + \dfrac{b}{n^2}d'$

where $n=$ the number of subperiods per cycle, and $d'=$ the deviation from the midpoint of the series measured in subperiods.

Appendix 3
Selected Normal curve and exponential values

1. Abridged Normal curve areas table

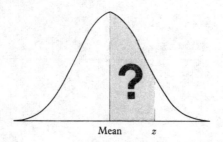

Mean z

z is the distance the point lies from the mean measured in σs, i.e.

$$z = \frac{Value - Mean}{\sigma}$$

(if z is minus, ignore sign).

z	Area	z	Area
0.0	0.0000	1.6	0.4452
0.1	0.0398	1.7	0.4554
0.2	0.0793	1.8	0.4641
0.3	0.1179	1.9	0.4713
0.4	0.1554	2.0	0.4772
0.5	0.1915	2.1	0.4821
0.6	0.2257	2.2	0.4861
0.7	0.2580	2.3	0.4893
0.8	0.2881	2.4	0.4918
0.9	0.3159	2.5	0.4938
1.0	0.3413	2.6	0.4953
1.1	0.3643	2.7	0.4965
1.2	0.3849	2.8	0.4974
1.3	0.4032	2.9	0.4981
1.4	0.4192	3.0	0.4987
1.5	0.4332		

Example

If a distribution has a mean of 30 and a σ of 5, what area lies under the curve between the mean and 38?

$$z = \frac{38 - 30}{5} = 1.6$$

The table shows that when $z = 1.6$ the area is 0.4452.
$\therefore$ Area lying under the curve = 44.52 per cent.

2. Negative exponential values table

x	e^{-x}	x	e^{-x}
0.1	0.9048	3.1	0.0450
0.2	0.8187	3.2	0.0408
0.3	0.7408	3.3	0.0369
0.4	0.6703	3.4	0.0334
0.5	0.6065	3.5	0.0302
0.6	0.5488	3.6	0.0273
0.7	0.4966	3.7	0.0247
0.8	0.4493	3.8	0.0224
0.9	0.4066	3.9	0.0202
1.0	0.3679	4.0	0.0183
1.1	0.3329	4.1	0.0166
1.2	0.3012	4.2	0.0150
1.3	0.2725	4.3	0.0136
1.4	0.2466	4.4	0.0123
1.5	0.2231	4.5	0.0111
1.6	0.2019	4.6	0.0101
1.7	0.1827	4.7	0.00910
1.8	0.1653	4.8	0.00823
1.9	0.1496	4.9	0.00745
2.0	0.1353	5.0	0.00674
2.1	0.1225	5.1	0.00610
2.2	0.1108	5.2	0.00552
2.3	0.1003	5.3	0.00499
2.4	0.0907	5.4	0.00452
2.5	0.0821	5.5	0.00409
2.6	0.0743	5.6	0.00370
2.7	0.0672	5.7	0.00335
2.8	0.0608	5.8	0.00303
2.9	0.0550	5.9	0.00274
3.0	0.0498	6.0	0.00248

Example $e^{-2.4} = 0.0907$

Appendix 4
Suggested answers to progress tests

Progress test 1

1. (*a*) Or kill them more quickly. Or send them home unfit. A shorter stay in hospital does not of itself mean a quicker cure.

(*b*) The average driver covers more miles on numerous short trips near to home than he does on journeys outside the 5-mile range. Under such circumstances, therefore, it is not surprising that more accidents occur within 5 miles of the driver's home than outside this range. Long journeys are not then necessarily safer, they are just less common.

(*c*) Let us assume that X is black coffee. Now it may well have been that the 10 per cent of the drivers involved in an accident had drunk black coffee because they had been feeling drowsy, and that, in fact, their accident was due to this drowsiness. Clearly, then, the black coffee was not a contributory cause of their accident. Indeed, the 1 per cent who took coffee and did not have an accident might well have done so if they had gone without – and if, say, 10,000 were *not* involved in an accident, then since 1 per cent of 10,000 is 100 this means that *not* taking black coffee could have resulted in double the number of accidents.

(*d*) How do you know? Did they tell you? And how many approached you that you did not save – and, indeed, how many did you perhaps drive to a life of crime they would not otherwise have contemplated?

(*e*) What is a good citizenship scale and how is it calibrated? What ranks as 'average'? How were the ratings obtained – by asking the boys? More important, because 'good citizens' tend to lead more stable lives they are easier to trace then 'bad citizens' who, of course, wish to be untraceable precisely because they are bad citizens. It is not surprising, then, that the majority of traceable boys prove to be good citizens.

(*f*) Far more people spend far more time in the home than in factories and mines. In these circumstances a 0.1 per cent accident rate in the home can easily result in more accidents than a 20 per cent rate in a dangerous factory (e.g. if 10,000 people are in the home, then 10 accidents will occur while if 40 people are in the factory only 8 accidents will occur). Also relevant to a measure of 'more dangerous' is the degree of hurt. One fatal accident is far worse than 20 cut fingers.

(*g*) This argument is perfectly valid and the conclusion perfectly sound.

Progress test 2

2.   280
     500
     641
     800
     900
     ___
     3,121 Answer = <u>3,100 tonnes</u>

Note: Since three figures are exact hundreds this implies that some figures are being rounded to the nearest 100. It is assumed, therefore, that 500, 800 and 900 are approximations, so the answer is approximated to the nearest 100 tons.

3. $1,200 \times 65 \times 4 = 312,000$.

Since the lowest number of significant figures in both the rounded figures used is two, the answer can only contain two significant figures.

∴ Total approximate weight of potatoes bought in a year
 = 310,000kg.

Note: The 4, being an exact number, is not a rounded number and therefore is excluded from the inspection for the number with the lowest number of significant figures.

4. (*a*) 21.388±0.056. (*b*) 21.332±0.056.

Progress test 3

3. *Comments.*

(*a*) This is not a random sample. Indeed, there is every probability of bias since there is a tendency for people living in the same neighbourhood to be physically similar. Thus, if one newsagent were just outside an old persons' home it would be quite possible for the observer to record a preponderance of white-haired and frail readers.

(*b*) The *buyer* of a magazine is not necessarily the *reader*. It is quite possible that some housewives will pick up their husbands' regular order while they are out shopping.

(*c*) Physical 'facts' collected in this way would not be very factual. Height and weight would need to be estimated and complexion would have to be judged subjectively.

The observers' results, then, might well conjure up in the mind of the reader a picture substantially different from the reality.

4. (*a*) Cluster sampling. (Random sampling is impossible as no sample frame exists here.)

(*b*) Multi-stage sampling and interviewing. (Interviewing will be necessary to obtain this type of personal information. The travelling involved as a result will render a random sample too expensive.)

(*c*) Stratified random sampling. (The number of West End, city, suburban and country cinemas can be found from official statistics.)

(*d*) Quota sampling.

5. Very often compilers of telephone directories adopt the principle of starting all their alphabetical sections with abbreviated names. This leads to the first entries under 'A' starting with names such as 'A.1 . . . Company' and 'A.B.C. . . . Company'. Even a business name beginning 'A.Z. . . .' will precede the name 'Aarons'. Added to this is the fact that the name 'Abbey . . .' is a favourite first name for many businesses. In consequence it will often be found that in many directories the first hundred or so names are predominantly business names. It follows, then, that any sample based on the first 200 names in an area telephone directory will be heavily biased towards commercial businesses.

Progress test 4

1. (Note that many of the figures for this table were, in fact, obtained from a short computer simulation.)

Road works queue lengths
Queue lengths with 2 different STOP/GO settings and 3 different arrival rates
Source: Field observations

Basic data			Arrival rate			
No. of vehicles:						
per minute	9		12		15	
per second	0.15		0.20		0.25	
	Cycle		Cycle		Cycle	
	100	160	100	160	100	160
Average vehicle arrivals per cycle	15	24	20	32	25	40
GO time (seconds)	30	60	30	60	30	60
STOP time (seconds)	70	100	70	100	70	100
Maximum throughput[1]	20	40	20	40	20	40
Observations						
Queue lengths as lights change to:						
GO: Average	10.86	14.50	34.42	20	n.a.	51.94
Maximum	23	24	59	31	n.a.	77
STOP: Average	0.37	0	20.58	0.23	n.a.	27.46
Maximum	8	0	48	5	n.a.	55
Multi-wait[2]	0.02	0	0.95	0	n.a.	0.83
Average vehicle waiting time (secs)	33	45	135	46	n.a.	170

1. Number of vehicles which can pass through the lights during the GO phase.
2. Proportion of occasions when the queue at the change to GO exceeds maximum throughput (i.e. when the queue is not cleared during the GO phase of vehicles which were in the queue during the earlier STOP phase).
n.a. Since the number of arrivals per cycle is greater than the maximum throughput the queue must inevitably lengthen each cycle and figures here are not therefore applicable to the engineer's study.

In this table the average vehicle waiting time has been estimated bearing the following considerations in mind:

(*a*) vehicles that arrive during the STOP phase and pass through on GO on average wait half the STOP time;

(*b*) vehicles that arrive during the GO phase and pass through at GO have no waiting time;

(*c*) vehicles that arrive during the STOP phase and fail to pass through at GO add the whole cycle time to their previous waiting time;

(*d*) vehicles that arrive during the GO phase and fail to pass through add half the GO time to the whole of the STOP time.

The following comments can be made about the table:

1. Setting a cycle time such that the maximum throughput is less than the arrival rate during the GO phase just leads to ever lengthening queues with no possible chance of the queue being cleared. Such cycle times, then, should be avoided as far as is humanly possible.

2. A cycle time set such that the maximum throughput equals the arrival rate (in the table, the 100 cycle with a 12-per-minute arrival rate and the 160 cycle with a 15-per-minute arrival rate) results in far longer queues than might have been expected if it were reasoned that since the two figures balance each other the queue would, on the majority of occasions, be all but cleared each GO phase.

3. The table suggests that there is a point of balance between arrival rates and maximum throughputs. A cycle time of 100 when the arrival rate was 9-per-minute (maximum throughput 20 and an arrival rate of 15 per GO phase) gave a shorter average queue waiting a shorter STOP period (albeit there was the very occasional time when one or two vehicles waiting during STOP failed to pass through on the following GO) than was given by a 160 cycle time for the same arrival rate. On the other hand, when the arrival rate rose to 12-per-minute (giving equal arrival and maximum throughputs in the case of the 100 cycle time) queue lengths were very much greater at the 100 cycle time than they were with the 160 cycle time.

4. It would seem that queues at traffic lights form and disperse in a more complex manner than might perhaps have been expected.

Notes on the construction of the table.

1. As the engineer clearly wanted to know which was the best cycle

time to fit the circumstances, and not which were the best circumstances to fit the cycle time, cycle times were put adjacent to each other – i.e. the arrival rate columns were divided into cycle times rather than the cycle time columns divided into arrival rates.

2. Because of the crucial relevance of certain figures which followed mathematically from the circumstances, these figures (and especially the very important maximum throughput figure) were incorporated into the table despite not being observed figures as such.

3. To cater for the mixture of observations and the figures referred to in 2, the table was divided horizontally into these two groups.

4. The 'Multi-wait' figure was not originally asked for by the engineer but since it would appear to be a useful figure in the context and since its computation was simply a matter of looking at the raw data and noting the proportion of times the observed queue length at the moment the lights changed to GO exceeded the maximum throughput, it has been included in the table.

2. *Criticisms of table.*

(*a*) No title.

(*b*) No source stated.

(*c*) No units given in *Weight of metal* column.

(*d*) (i) What are *Foundry hours*?

   (*ii*) Since all weights of castings are covered by the first three lines of the table, what can 'Others' (line 4) refer to?

   (*iii*) Does 'Up to 4kg' include or exclude a casting weighing 4kg? Similarly, does 'Up to 10kg' include or exclude a casting weighing 10kg?

(All these are examples of ambiguity.)

(*e*) Since 'Up to 10kg' includes 'Up to 4kg' it would seem that double-counting is occurring.

(*f*) The *Foundry hours* do not add up to 2,000. What is this latter figure, then? If this total is relevant to the table, what is missing from the main body of it?

Summary. A confused table which, at best, tells little or nothing and, at worst, could mislead.

Progress test 5

1. *See* Fig. A4.1.

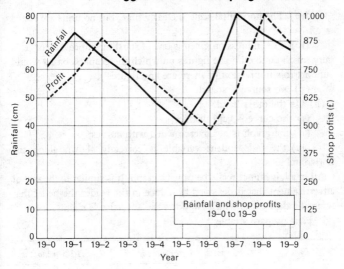

Figure A4.1 *Rainfall and shop profits.*

Note: To answer the question it was necessary to use a double scale, and in order to bring out the relationship in the most effective manner the two scales were chosen so that the two curves occupied the same part of the graph.

As a result of such a choice of scales it becomes obvious from the graph that the two curves are almost identical, the main difference being that the rainfall curve *precedes* the profit curve by a year. This suggests that profits are closely related to the *Previous* year's rainfall. (In statistics, when one curve follows another the time difference is termed *lag*. Finding cases of lag in times series is often useful since it means that the future value of one variable can be closely estimated from the current value of another – *see* 12:**2**.)

2. (*a*) This graph has the following *serious* faults.

(*i*) The title is not clear. What is 'improvement' and how is it measured?

(*ii*) The vertical axis shows neither heading nor units (it is presumed to be some sort of 'improvement' scale – if, in fact, there can be such a scale). Moreover, it may not even start at zero.

(*iii*) The horizontal scale is clearly time, but no units are stated at all.

(*iv*) The impression given is one of startling improvement. It is very much doubted if the figures on which the graph is based would support this impression (if there are any figures!).

(*v*) No source is stated.

(*b*) On the other hand:

(*i*) the curve is very distinct;

(*ii*) the graph is not overcrowded with curves;

(*iii*) the independent variable is correctly shown along the horizontal axis.

All in all, it cannot really be called a graph. It is similar to the sort of advertisement that is designed to induce in the reader the belief that scientific data exists which supports the advertiser's claims.

3. First prepare a table:

Year	Amount	3-year moving Total[1]	3-year moving Average	10-year moving Total[1]	10-year moving Average
1960	5				
1961	8		$6.\dot{3}$[2]		
1962	6	19	$8.\dot{6}$[3]		
1963	12	26	$7.\dot{3}$		
1964	4	22	8.0		
1965	8	24	9.0		
1966	15	27	11.0		9.1[4]
1967	10	33	$11.\dot{6}$		10.6
1968	10	35	11.0		11.4
1969	13	33	$14.\dot{3}$	91	12.3
1970	20	43	$16.\dot{3}$	106	11.7
1971	16	49	17.0	114	13.1
1972	15	51	$12.\dot{3}$	123	13.2
1973	6	37	13.0	117	13.2
1974	18	39	11.0	131	13.0
1975	9	33	14.0	132	13.2
1976	15	42	$10.\dot{6}$	132	13.3
1977	8	32	$11.\dot{6}$	130	13.1
1978	12	35	$11.\dot{3}$	132	13.7
1979	14	34	$14.\dot{6}$	133	13.8
1980	18	44	18.0	131	14.6
1981	22	54	$18.\dot{6}$	137	14.8

1982	16	56	17.3	138
1983	14	52	16.6	146
1984	20	50		148

Now *see* Fig. A4.2.

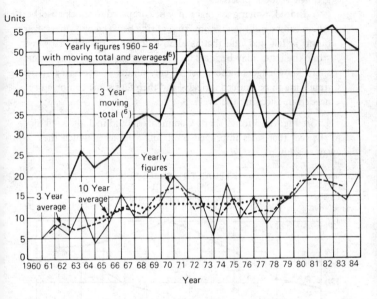

Figure A4.2 *Yearly figures with moving total and averages*

Difference between 10-year and 3-year moving averages. The 10-year moving average smooths out the fluctuations far more than the 3-year average; in fact, the 10-year average is nearly a straight line. However, the 3-year average is more sensitive to changes and signals a new trend sooner than the 10-year which tends to lag behind.[7]

Notes:
   1. To calculate a moving average it is first necessary to calculate a moving total.
   2. This average is located at the midpoint of its 3-year period.
   3. A dot over a decimal digit means that the decimal is recurring.
   4. This average is located at the midpoint of its 10-year period.

5. Totals on this graph are plotted at the *end* of the period to which they apply, and averages at the *midpoint* (see 5:**14**(*a*)(*ii*)).

6. A moving total of 3-years is really too short a period to give a curve of any significance. It has been included simply in order to give the student practice.

7. Indeed, moving averages may be compared to shock-absorbers which can be built to give any kind of ride between one where every bump is felt and one that is almost perfectly smooth. However, while the latter may be ideal for riding, in economics and business some knowledge of the most recent bumps is necessary if new trends are to be noticed quickly, and therefore some compromise is needed in choosing the average.

4. *See* Fig. A4.3 for Z chart. Comments should include a reference to the fact that sales improved until July 19-6 after which they fell away steadily.

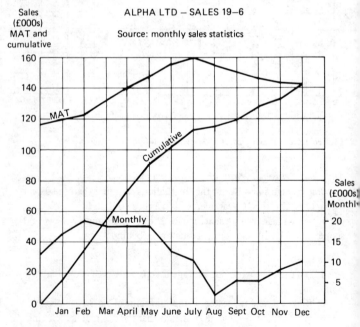

Figure A4.3 *Z chart*

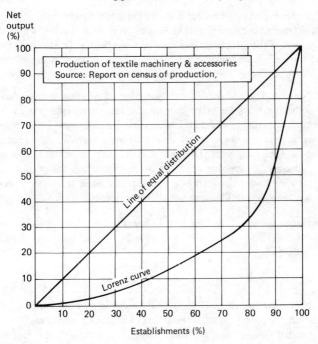

Figure A4.4 *Lorenz curve of production of textile machinery and accessories.*
The following data is relevant.

Establishments			Net output		
No.	%	Cum.%	£000s	%	Cum.%
48	22.5	22.5	1,406	3.5	3.5
42	19.5	42	2,263	6	9.5
38	18	60	3,699	9.5	19
26★	12	72	3,152★	8	27
21	10	82	2,836	7.5	34.5
16	7.5	89.5	5,032	13	47.5
23	10.5	100	20,385	52.5	100
214			38,773		

★ *See* note on following page.

5. *See* Fig. A4.4. This curve shows the extent to which the sharing of net output between establishments diverges from equality. Thus 82 per cent of the establishments have between them only 34½ per cent of the net output, or, put the other way round, 18 per cent of the establishments are responsible for over 65 per cent of the net output.

Note to Fig. A4.4.★ In the question this figure and the one following were the other way round. The interchange was made because to obtain a smooth Lorenz curve it is necessary for *the order of the figures to run from those establishments having the least output to those having the most*. To check that the order is correct it is only necessary to see that the *average* continually increases. In the question layout the average dropped at this point and so the order was rearranged. Thus the layout in the question gave:

Establishments	Output	Average
38	3,699	97½
21	2,836	135
26	3,152	121
16	5,032	314½

Normally, students will find that Lorenz curves are constructed from frequency distributions (*see* Table 5C) and, since the construction of such distributions involves arranging the data in order from the lowest to the highest, interchanges of this sort are rarely necessary.

6. (*a*) The graph shows that unemployment in Fredsville rose very sharply between 31 December 19–1 and 31 December 19–3, and then fell quite rapidly, although not as rapidly as it had risen, over the next three years. However, the rise and fall were not experienced equally by the four groups of unemployed. Up to the peak the greatest increase in unemployment was among the shorter-term unemployed – those unemployed under 12 weeks – although the pro rata increase was much the same for all four groups. In the fall from the peak the differences between the groups was even more. Those unemployed less than three weeks quickly fell in number until 19–6 when very many fewer were unemployed than in 19–1. Much the same happened to those unemployed from 3 to under 12 weeks, although by 19–6 unemployment was still as high as in 19–1. For

those unemployed between 12 weeks and a year the position hardly changed – for these there was no return to 19–1. And the long-term unemployed suffered the most for their unemployment continued to increase even after the peak unemployment was over. In fact, the rate of increase grew until 19–5 and only after that did the numbers stabilise.

It should, of course, be appreciated that the actual people comprising an unemployed group in a particular year would rarely be the same people the following year. Thus, someone unemployed in 19–3 for under three weeks may well not have seen his chance of gaining employment grow at all, for he would move out of the 'under 3 weeks' category very quickly. Indeed, had he remained unemployed until 19–4 he would have appeared in the 'over 1 year' category. Although the chart figures comprise actual individuals, the chart says nothing about the unemployment of specified individuals. For such a chart you would need to start with all the unemployed at 31 December 19–1 and then record how many of these remained unemployed year by year. The resulting figures would then give a single curve quite unlike our band chart.

(b) If the figures had not been contrived it would have been virtually impossible for each of the six years' unemployment figures to have been an exact hundred, let alone the unemployment in each category a multiple of ten.

It is important to form the habit of looking at data critically. It is not, perhaps, so much a matter of figures being 'cooked' (the case in this exercise) as that there may be some undeclared feature of the method of collection which makes the data appear different than it really is. Frequently this may prove to be the rounding of data although it could, for instance, be the disregard of the particular category as when a figure relating to exports disregards (but fails to state this disregard) that re-exports are excluded or that export consignments of under £100 are ignored. Critical appraisal of the data can often prevent the user being misled in this way.

Progress test 7

1.

(a) Age (years)	(b) Extractions	(c) Income p.a. (£)
20–under 25	3–5[2]	Under £2,550[4]
25–under 30	6–8	2,550– 2,849[5]
30–under 35	9–11	2,850– 3,149
35–under 40	12–14	3,150– 3,449
40–under 45	15–17	3,450– 3,749
45–under 50	18–20	3,750– 4,249[6]
50–under 55	21–23	4,250– 4,749
55–under 60	24–26	4,750– 5,249
60–under 65	27–29	5,250– 5,749
65–under 70	30–32	5,750– 6,249
70–under 75	33–34	6,250– 6,749
75–under 80		6,750– 7,249
80–under 85	Class limits '6–8':	7,250– 7,749
85–under 90	5½ to 8½[3]	7,750– 9,249
90–under 95	extractions	9,250–10,749
95–under 100		10,750–14,249
100 and over		14,250–19,749
		19,750–30,249
Class limits of '25–		30,250–44,999
under 30': 25 years		45,000 and over[4]
exactly up to, but		Class '2,550–2,849':
not including, 30		£2,549½ to £2,849½[7]
years[1]		

Notes:

1. It should always be remembered that ages are usually given as at the last birthday, i.e. rounded down. Any distribution using ages should be constructed with this in mind. In fact, students are warned that official figures often reflect this by showing *stated* limits as 20–24; 25–29, etc., and so on.

2. A class interval of 3 is not usually recommended, but it was chosen here as a compromise. An interval of 2 would be so small that one might as well construct the distribution without any grouping at all and so avoid the loss of information that inevitably accompanies grouping. On the other hand, an interval of 4 would result in only 7 or 8 classes.

With 3 as an interval, the distribution 'looks' better if the stated lower class limits are multiples of 3.

3. The data is discrete – you cannot have half an extraction – so mathematical limits extending to half a unit on each side of the stated limits are used.

4. These classes are open-ended. It is assumed that few full-time employed adults have incomes below £2,550 p.a. or over £45,000 p.a.

5. This first group of classes has been chosen so that the round hundreds lie symmetrically throughout the class, as it is assumed that there will be a tendency for incomes to be set at the round hundred level. This arrangement is in compliance with 7:**14**(*d*) (note, however, that incomes set at the round £50 level will unfortunately result in some undesirable clustering at the beginning of classes).

6. Henceforth limits which ensure that round £1,000s and £500s lie symmetrically throughout the class are selected. Also classes are unequal for reasons given in 7:**12**.

7. Clearly, if the data collected was recorded to the nearest £ then the *true* limits extend £½ out from the stated limits.

2. (*a*) *Kilometres recorded by 120 salesmen in the course of one week*

Kilometres	Frequency (*f*)
390–under 410	4
410–under 430	17
430–under 450	33
450–under 470	31
470–under 490	20
490–under 510	13
510–under 530	2
	120

(*b*) 429½–under 449½ km.

Progress test 8

1. Set out the distribution:

Class	f	Cumulative f
0–under 10	12	12
10–under 25	25	37
25–under 40	51	88
40–under 50	48	136
50–under 60	46	182
60–under 80	54	236
80 and over	8	244

After laying out this distribution the first thing to decide, in view of the unequal class intervals, is what unit of density should be chosen for the histogram. Since three of the classes have intervals of '10' let us say that a strip 10 units long and 1 unit high represents a frequency of '1'. This means that the classes '0–under 10', '40–under 50' and '50–under 60' will have rectangles 12, 48, and 46 units high. The other classes will have to have their heights adjusted by dividing their frequencies by one-tenth their class intervals, i.e.:

Class	Class interval	Divide f by:	Rectangle height (units)
10–under 25	15	1.5	16⅔
25–under 40	15	1.5	34
60–under 80	20	2	27
80–under 100	20	2	4

We can now construct the required graph: *see* Fig. A4.5.

Notes to Fig. A4.5.

1. In plotting the adjusted frequencies the original frequencies do not show on the graph. Under these circumstances it is normal to write the original frequencies over the rectangles.

2. Note that in the construction of ogives unequal class intervals do not lead to the sort of special adjustments needed when constructing histograms.

2. (*a*) 85 distances.
   (*b*) 400 to 432 kilometres.

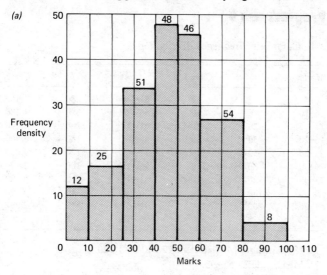

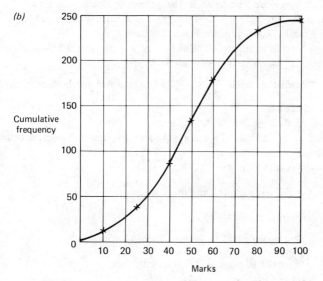

Figure A4.5 (a) *Histogram of marks*[1]. (b) *Ogive of marks* [2]

Progress test 9

1. Cumulative frequency distribution:

Marks	f	Cum. f	Marks	f	Cum. f
0–5	2	2	51– 55	280	1,160
6–10	8	10	56– 60	320	1,480
11–15	20	30	61– 65	260	1,740
16–20	30	60	66– 70	160	1,900
21–25	50	110	71– 75	120	2,020
26–30	80	190	76– 80	100	2,120
31–35	120	310	81– 85	60	2,180
36–40	150	460	86– 90	40	2,220
41–45	200	660	91– 95	17	2,237
46–50	220	880	96–100	3	2,240

(a) (i) *Median*: Median item = 2,240/2 = 1,120.

∴ Median class = 51–55.

Now there are 880 items in the preceding classes. Therefore the median item is the 1,120 − 880 = 240th item beyond the beginning of the median class, which contains 280 items.

∴ Median item lies 240/280ths of the way through the class. Now the class has an interval of 5 marks.*

∴ Median value is $\frac{240}{280} \times 5 = 4.29$ marks above the bottom of the median class.

∴ Median = 4.29 + 50.5* = 54.79

But as no fractions of a mark were awarded the median must be a round number.

∴ Median = 55 marks.

(ii) *First quartile*: First quartile item = $\frac{2,240}{4}$ = 560th

∴ Quartile class = 41–45

∴ First quartile = $40.5 + \frac{560 - 460}{200} \times 5 = \underline{43 \text{ marks.}}$

(iii) *Third quartile*: Third quartile item

$= \frac{2,240}{4} \times 3 = 1,680$th

$\therefore$ Quartile class = 61–65

$\therefore$ Third quartile = $60.5 + \dfrac{1,680 - 1,480}{260} \times 5 = 64.35$

i.e. rounded to the nearest unit = <u>64 marks.</u>

(*iv*) *Sixth decile:* Sixth decile item = $\dfrac{2,240}{10} \times 6 = 1,344$th

$\therefore$ Sixth decile class = 56–60

$\therefore$ Sixth decile = $55.5 + \dfrac{1,344 - 1,160}{320} \times 5 = 58.37$

i.e. to the nearest unit = <u>58 marks.</u>

(*v*) *42nd percentile:* 42nd percentile item
$$= \dfrac{2,240}{100} \times 42 = 940.8$$

i.e. since there can be no fraction of an item, 941st item.

$\therefore$ 42nd percentile class = 51–55

$\therefore$ 42nd percentile = $50.5 + \dfrac{941 - 880}{280} \times 5 = 51.59$

i.e. to nearest unit = <u>52 marks.</u>

(*b*) If the examining body wished to pass only one-third of the candidates, the cut-off mark would need to be the mark obtained by the candidate one-third from the top, i.e. the $2,240 - \frac{1}{3} \times 2,240 = 1,493$rd candidate.

This candidate's marks fall in the 61–65 class and can be estimated as

$$60.5 + \dfrac{1,493 - 1,480}{260} \times 5 = 60.75$$

This means that the pass mark must be set at <u>61 marks.</u>

Note: *Since no fractions of a mark are given, the data is discrete and so the class limits are the mathematical limits and are extended, therefore, a half mark either side of the stated limits, e.g. class 51–55 is considered to have limits of 50.5–55.5. This means that, first, the class interval is 5 and, second, that the lower class limit is 50.5.

2. (a)

Earnings (£)	f	Cum f
<3000	7	7
3000– <4000	30	37
4000– <5000	37	74
5000– <6000	51	125
6000– <7000	32	157
7000– <8000	25	182
8000–<10000	14	196
>10000	4	200

The ogive constructed from these figures is shown in Fig. A4.6.

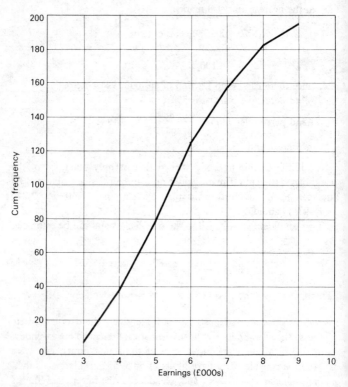

Figure A4.6 *Ogive of earnings*

Note that since the upper and lower classes are open-ended, these two classes cannot be shown on the ogive as there is no lower limit of the lower class or upper limit of the upper class which have values to be plotted in conjunction with the cumulative frequencies of 0 and 200 respectively.

(b)

Measure	Fractile item	Fractile (£)*
Median	100th	5410
Upper quartile	150th	6780
Lower quartile	50th	4350
Highest decile	180th	7920

*For the purpose of providing a suggested answer, these figures have, in fact, been computed.

Progress test 10

1. (a) Ungrouped frequency distribution of the data:

No. of passengers		f	Cumulative f
135	‖‖	4	4
136	‖‖‖ ‖‖	8	12
137	‖‖‖	5	17
138	‖‖	3	20
		$\Sigma f = 20$	

No. (x)	f	fx*
135	4	540
136	8	1,088
137	5	685
138	3	414
	n = 20	$\Sigma fx = 2,727$

$$\therefore \text{Mean } (\bar{x}) = \frac{2,727}{20} = \underline{136.35 \text{ passengers.}}$$

(*i*) The most frequently recurring value is 136.

∴ Mode = <u>136 passengers.</u>

(*ii*) The middle items of this array are the 10th and 11th. The cumulative frequency distribution shows that the values of both these items are the same, i.e. 136.

∴ Median = <u>136 passengers.</u>

(*b*) If the last item were 35 instead of 135 then Σfx would be 100 less, i.e. 2,627.

∴ Mean $(\bar{x}) = \dfrac{2,627}{20} = \underline{131.35 \text{ passengers.}}$

This correction of the last item recorded simply means that one of the early items in the array has a value less than before, but this makes no difference to the values of the 10th and 11th items (though note that if it had been one of the *later* items in the array the 10th and 11th items would be shifted one down from where they previously were). Therefore the median remains unaltered. Similarly the mode remains unaltered since 136 still occurs most frequently.

The revised figures show that the mean can be greatly distorted by an extreme value. The revised mean now stands well below *any of the other 19 values in the distribution*. On the other hand the median and mode remain unaltered and this demonstrates that these two averages are not affected by extreme values.

Notes on answer

*If a value is multiplied by the number of times it occurs, i.e. fx, the product is clearly equal to the figure obtained if all the occurrences of that value were added, e.g. $4 \times 135 = 135 + 135 + 135 + 135$. Therefore Σfx = the sum of all the values in the distribution.

2. (*a*) School A:

IQ (x)	f	MP	f×MP	Cum. f
75–under 85	15	80	1,200	15
85–under 95	25	90	2,250	40
95–under 105	40	100	4,000	80
105–under 115	108	110	11,880	188
115–under 125	92	120	11,040	280
125–under 135*	20	130	2,600	300
Total	300		32,970	

Mean IQ: $\bar{x} = 32,970/300 = \underline{109.9 \text{ IQ marks.}}$

Median IQ: Median item = 300/2 = 150th item.

∴ Median class is 105–under 115

Now the median item is 150 − 80 = 70th item in a class of 108 items.

∴ The median item lies 70/108th into the class 105–under 115. And since the class interval is 10 units, then 70/108th of the interval = $(70/108) \times 10 = 6.5$ units.

∴ Median lies 6.5 units above the bottom of the 105–under 115 class, i.e. 105 + 6.5 = <u>111.5 IQ marks.</u>

School B:

Similarly, mean = <u>95.3 IQ marks.</u>

median = <u>93.3 IQ marks.</u>

*For the purpose of computing the mean an open-ended class is assumed to be the same size as the adjoining class (7:**13**).

(*b*)

School	Mean IQ	No. of pupils	Total IQ points in school
A	109.9	300	32,970
B	95.3	250	23,825
C	106.0	450	47,700
		Σw 1,000	Σxw = 104,495

∴ Weighted average (mean IQ of all pupils in the three schools)
= 104,495/1,000 = 104.495
≅ 104.5 IQ marks.

3. Let f_s = frequency of science classes and f_a = frequency of arts classes.

(*a*)

No. students	f_s	f_a	Class MP†	Total students* $f_s \times MP$	$f_a \times MP$
1–6	4	0	$3\frac{1}{2}$	14	0
7–12	15	3	$9\frac{1}{2}$	$142\frac{1}{2}$	$28\frac{1}{2}$
Subtotal (<13)	19	3		$156\frac{1}{2}$	$28\frac{1}{2}$
13–18	11	10	$15\frac{1}{2}$	$170\frac{1}{2}$	155
19–24	8	8	$21\frac{1}{2}$	172	172
25–30	5	4	$27\frac{1}{2}$	$137\frac{1}{2}$	110
31–36	1	1	$33\frac{1}{2}$	$33\frac{1}{2}$	$33\frac{1}{2}$
Σ	44	26		670	499

*Note that if we multiply the class frequency by the class midpoint (average number of students in the class) we obtain the total number of students in the class.

†As the data is discrete mathematical limits of $\frac{1}{2}$–$6\frac{1}{2}$, $6\frac{1}{2}$–$12\frac{1}{2}$, etc. are used.

$$\text{So mean class size} = \frac{670+499}{44+26} = 16.7 \text{ students}$$

(b) (i) The total students are the same. The number of classes are now $(f_s - \text{subtotal} <13) + (f_a - \text{subtotal} <13) = (44-19) + (26-3) = 48$

$$\text{So mean class size} = \frac{670+499}{48} = 24.35 \text{ students}$$

(ii) The number of classes is as in (i). The total number of students is now $(\Sigma f_s \times \text{MP} - \text{subtotal} <13) + (\Sigma f_a \times \text{MP} - \text{subtotal} <13)$
$= (670 - 156\frac{1}{2}) + (499 - 28\frac{1}{2}) = 984$

So mean class size $= 984/48 = \underline{20.5 \text{ students}}$

(c) 1980–81 enrolments: $670 + 20\%$ and $499 - 10\% = 804 + 449$
$= 1253$ students

Now $\bar{x} = \Sigma x/n$. So if $\bar{x}$ is not to be less than 20 then:

$20 = 1253/n$, i.e. $n = 1253/20 = 62.67$ classes

Since there must be a discrete number of classes, then if $\bar{x}$ is not to be less than 20, the maximum number of classes must be $\underline{\underline{62}}$

Note: There is a slightly shorter solution that involves weighted averages. Can you find (have you found) it?

4. (a) Cumulative frequency distribution:

Length of wait (min)	f	Cumulative f
0	50	50
Over 0–under 0.5	210	260
0.5–under 1	340	600 (First quartile class)
1–under 2	200	800 (Median class)
2–under 3	110	910
3–under 5	170	1,080 (Third quartile class)
5–under 10	140	1,220
10 and over	80	1,300

The middle 50 per cent of the customers are those who lie between the first and third quartile (*see* 9:**6**).

First quartile item = 1,300/4 = 325th item

∴ First quartile

$$= 0.5 + \frac{325 - 260}{340} \times 0.5 = 0.096 = \underline{0.596}$$

Third quartile item = 1,300/4 × 3 = 975th item

∴ Third quartile $= 3 + \frac{975 - 910}{170} \times 2 = 3 + 0.765 = \underline{3.765}$

∴ The middle 50 per cent of the customers wait between 0.596 and 3.765 minutes, i.e. between 36 seconds and 3 minutes 46 seconds.

(*b*) (*i*)

Length of wait (min)	Midpoint	f	f × midpoint
0	0	50	0
Over 0–under 0.5	0.25	210	52.5
0.5–under 1	0.75	340	255.0
1–under 2	1.5	200	300.0
2–under 3	2.5	110	275.0
3–under 5	4	170	680.0
5–under 10	7.5	140	1,050.0
10 and over	12.5	80	1,000.0
Total		1,300	3,612.5

So $\bar{x} = \frac{3,612.5}{1,300} = \underline{2.78 \text{ minutes}}$

(*ii*) From the cumulative frequency distribution drawn up in (*a*) it is possible to determine the median.

Median item = 1,300/2 = 650th

∴ Median $= 1 + \frac{650 - 600}{200} \times 1 = 1 + 0.25 = \underline{1.25 \text{ minutes.}}$

This means that half the customers wait 1.25 minutes or less and the other half 1.25 minutes or more. Now the chances are equal as to which half you will be in[*], and therefore you will have a 50/50 chance of being at the checkpoint in 1.25 minutes or less. Since you wish to be there in 2 minutes, the odds are definitely in your favour.[†]

Notes:

*Look at it this way: *all* customers must fall into one half or the other. You are one of the customers. Since there is nothing to make you fall in one half more than another, the chance of being in the first half is the same as the chance of being in the second half.

†Students are sometimes puzzled as to why, if your mean wait is 2.78 minutes, you will have a good chance of reaching the checkout point in under 2 minutes. The reason is this:

The 2.78 minutes is the mean waiting time over a large number of visits. Now on some of these visits the waiting time was over 10 minutes. As has been pointed out in this chapter, extreme values tend to distort the mean, and consequently these excessive waits make the mean waiting time longer than the majority of actual waiting times (even with the mean at 2.78 minutes, one wait of 10 minutes would need to be balanced by virtually three visits with no waiting time at all if the mean were to stay unchanged).

5. (a)

Machine	No. of minutes per article	No. of articles per hour
A	2	$60/2 = 30$
B	3	$60/3 = 20$
C	5	$60/5 = 12$
D	6	$60/6 = \underline{10}$
Total articles per hour =		$\underline{\underline{72}}$

(b)

Machine	Production		Total
	In 1st and 2nd hour	In 3rd hour	
A	0	30	30
B	40	20	60
C	24	0	24
D	20	10	$\underline{30}$
			$\underline{\underline{144}}$

∴ 144 articles were produced in the 3 hours.

∴ Average number of articles per hour

$= 144/3 = \underline{48 \text{ articles}}$.

6. (a) *The arithmetic mean.* The few high wages will result in this mean being the highest of the averages. Although such extreme wage payments would be unlikely to distort the mean seriously, their effect would result in the average looking larger than the figures in the whole of the distribution would warrant.

(b) *The mode.* In this sort of distribution the most frequently occurring wage is inevitably near the bottom, and choice of this average would result in a figure which made no allowance for the few high wages that some members of the foremen's union would be earning.

(c) *The median.* The median wage indicates the wage of the foreman who is 'half way up the ladder'. Half the foremen do better and half do worse. So you have a 50/50 chance of earning the median wage or better.

7. A weighted average is found by multiplying the number of items in each group by the respective means of each group; adding the products; and finally dividing by the total number of items. As a formula this can be written so:

$$\text{Combined mean} = \frac{n_1 \bar{x}_1 + n_2 \bar{x}_2 + n_3 \bar{x}_3}{n_1 + n_2 + n_3}$$

Progress test 11

1. *School A*

Note: In Progress test 10, question 2(a), $\bar{x}$ was found to be 109.9 IQ marks.

IQ (x)	f	MP	(MP−$\bar{x}$)	f(MP−$\bar{x}$)	MP^2	$f \times MP^2$
75–under 85	15	80	29.9	448.5	6,400	96,000
85–under 95	25	90	19.9	497.5	8,100	202,500
95–under 105	40	100	9.9	396.0	10,000	400,000
105–under 115	108	110	0.1	10.8	12,100	1,306,800
115–under 125	92	120	10.1	929.2	14,400	1,324,800
125 and over	20	130	20.1	402.0	16,900	338,000
Total	300			2,684.0		3,668,100

Mean deviation:

$$\text{Mean deviation} = \frac{\Sigma f(\text{MP} - \bar{x})}{\Sigma f} = \frac{2,684}{300} = \text{approximately } \underline{9 \text{ IQ marks}}$$

Standard deviation:

$$s = \sqrt{\frac{\Sigma(f \times MP^2)}{\Sigma f} - \overline{x}^2} = \sqrt{3,668,000/300 - 109.9^2}$$
$$= \underline{12.21 \text{ IQ marks}}$$

Variability:

Coefficient of variation $= \frac{s}{\overline{x}} \times 100 = \underline{11.1}$

School B

Similarly (workings not shown), mean deviation $= 9\frac{1}{2}$ IQ marks, $s = 11.2$ IQ marks and coefficient of variation $= 11.75$.

Therefore the variability of the IQs in School B is slightly greater than of School A.

2. (a) $Q_1 = £2.884$ and $Q_3 = £4.252$

$\therefore$ Quartile deviation $= \frac{4.252 - 2.884}{2} = \underline{68.4p}$

(b) $\overline{x} = £3.5$. Mean deviation $= £0.708$. Standard deviation $= £0.9653$.

Sales (£000)	f	MP	f×MP	MP²	f×MP²	Cum. f
Under 10	25	5	125	25	625	25
10–under 20	18	15	270	225	4,050	43
20–under 30	8	25	200	625	5,000	51
30–under 40	3	35	105	1,225	3,675	54
40 and over	1	45	45	2,025	2,025	55
Total	55		745		15,375	

$\therefore \overline{x} = 745/55 = 13.55$, i.e. mean is $\underline{£13,550}$

$s = \sqrt{15,375/55 - 13.55^2} = 9.8$, i.e. standard deviation is $\underline{£9,800}$

Median item is the 28th.

$\therefore$ median $= 10 + \frac{3}{18} \times 10 = 11.67$, i.e. the median is $\underline{£11,670}$

So coefficient of skewness $= \frac{3(13,550 - 11,670)}{9,800} = \underline{+0.576}$

4. $Sk = \frac{3(20 - 22)}{10} = \underline{-0.6}$

Note that this distribution is *negatively* skewed.

5. *Estimate: £5,600.*

In a symmetrical distribution the mean would be the same as the median. However, the fact that the upper quartile (£6,780) is further away from the median (£5,410) than the lower quartile (£4,350) – i.e. £1,370 away as against £1,060 – indicates that the distribution is positively skewed. So the mean, then, will be higher than the median. To allow for this the middle of the interquartile range (£4,350 to £6,780), rounded to the nearest 100, has been taken as an estimate of the mean – which gives £5,600.

Note: Calculations show the mean, in fact, is £5,675.

6. This is a typical unstatistical statement. The following points arise in connection with the remark:

(*a*) Average what? All averages are measured and therefore have units. The comment makes no reference to units. Are the children alleged to be below average as regards height, weight, intelligence, scholastic ability, welding skills or skulduggery? As the statement stands it is meaningless.

(*b*) Below whose average? It is important that statements of this sort ensure that the measure is a measure of the population remarked upon. Thus, if the 'average' referred to was adult height then it is not surprising that half our children are below the average adult height – indeed, the converse would be more likely to warrant comment. But quite apart from such a crude divergence of average from population remarked upon, more subtle divergences can occur. So it would be possible to say that in a particular area two-thirds of the workers (including part-time workers) earn less than the average wages (of all full-time employees).

(*c*) Which average? If the mean is being referred to and a distribution is positively skewed then it follows that over one half will be less than the average (since the median in such a distribution lies below the mean). If it is the mode that is being referred to and the distribution is negatively skewed then it again follows that over half will be below the average (since the median in such a distribution is below the mode). If it is the median that is being referred to then the statement is nonsense since by definition no more than half the distribution can lie below the median.

(*d*) Who are 'our' children? If it is the children of the speaker who are being referred to and the measure is statistical understanding and statistical understanding is hereditary, then the remark is hardly likely to come as a surprise to anyone.

Answers to Assignment 3
The measures computed from the full distribution are as follows:
Mean, 51.039. Median, 36.033. Mode, 8.160. Quartile deviation, 29.747. Mean deviation, 34.191. Standard deviation, 41.255. Coefficient of variation, 80.830. P coefficient of skewness, 1.091.

Progress test 12

1. *See* Fig. A4.7. Note that IQ is the independent variable and is therefore plotted on the horizontal axis.

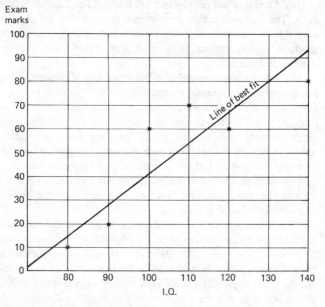

Figure A4.7 *Scattergraph of data in Progress test 12, question 1*

(*a*) IQ 130: estimated mark 80.
(*b*) Mark 77: estimated IQ 127.

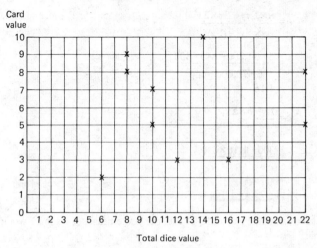

Figure A4.8 *Scattergraph of data in Progress test 12, question 2*

2. *See* Fig. A4.8. It is not possible to draw a line of best fit on this graph. Therefore the variables are not related.

Progress test 13

1. (*a*) *Regression line of examination marks on IQ*

$y = a + bx$, where y = examination marks and x = IQ

x	y	xy	x^2
110	70	7,700	12,100
100	60	6,000	10,000
140	80	11,200	19,600
120	60	7,200	14,400
80	10	800	6,400
90	20	1,800	8,100
Σs: 640	300	34,700	70,600

and $n = 6$

$$b = \frac{n \times \Sigma xy - \Sigma x \times \Sigma y}{n \times \Sigma x^2 - (\Sigma x)^2}$$

$$= \frac{6 \times 34,700 - 640 \times 300}{6 \times 70,600 - 640^2}$$

$$= 1.16$$

$$a = \frac{\Sigma y - b \times \Sigma x}{n}$$

$$= \frac{300 - 1.16 \times 640}{6}$$

$$= -73.4$$

$$\therefore y = \underline{-73.4 + 1.16x}$$

If a candidate had an IQ of 130, the best estimate of his mark would be:

$$y = 1.16 \times 130 - 73.4 = \underline{77 \text{ marks.}}$$

(since fractional marks do not appear to be awarded)

(b) *Regression line of IQ on marks*
$x = a + by\star$

$$\Sigma y^2 = 70^2 + 60^2 + 80^2 + 60^2 + 10^2 + 20^2 = 19,000$$

$$\therefore b = \frac{6 \times 34,700 - 640 \times 300}{6 \times 19,000 - 300^2} = 0.675$$

$$\therefore a = \frac{640 - 0.675 \times 300}{6} = 72.92$$

So $x = \underline{72.92 + 0.675y}$

$\therefore$ Best estimate of the IQ of a candidate who obtained 77 marks:

$$x(\text{IQ}) = 72.92 + 0.675 \times 77 = \underline{125}$$

See Fig. A4.9 for the superimposition of these lines on the scattergraph in Fig. A4.7.

Note: $\star$The a and b in this equation are, of course, quite different from the a and b of the equation in part (a) of the answer (though the xs and ys relate to the same variables).

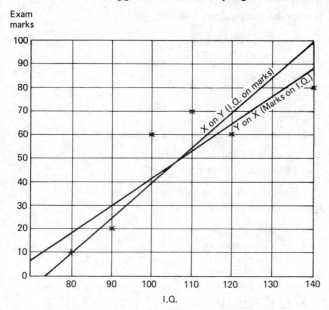

Figure A4.9 *Regression lines of data in Progress test 12, question 1*

2. Let x = die value, and y = card value, where n = 10.

x	y	x^2	y^2	xy
8	8	64	64	64
8	9	64	81	72
14	10	196	100	140
22	5	484	25	110
22	8	484	64	176
16	3	256	9	48
12	3	144	9	36
6	2	36	4	12
10	7	100	49	70
10	5	100	25	50
Σs: 128	60	1,928	430	778

(a) *Regression line of card value on die value*

Appropriate line equation: $y = a + bx$.

$$b = \frac{10 \times 778 - 128 \times 60}{10 \times 1928 - 128^2} = 0.0345$$

$$a = \frac{60 - 0.0345 \times 128}{10} = 5.56$$

So $y = 5.56 + 0.0345x$

$\therefore$ Estimate of card value y when die value x is:

(*i*) 4:$y = 5.56 + 0.0345 \times 4 = 5.70\star$
(*ii*) 24:$y = 5.56 + 0.0345 \times 24 = 6.39\star$

(*b*) *Regression line of die value on card value*
Appropriate line equation: $x = a + by$

$$b = \frac{10 \times 778 - 128 \times 60}{10 \times 430 - 60^2} = 0.14$$

$$a = \frac{128 - 0.14 \times 60}{10} = 11.94$$

So $x = 11.94 + 0.14y$

$\therefore$ Estimate of die value x when card value y is:
(*i*)   1:$x = 11.94 + 0.14 \times 1\ \ = \underline{12.08}\star$
(*ii*) 10:$x = 11.94 + 0.14 \times 10 = \underline{13.34}\star$

Note: $\star$In these examples it is not possible, of course, for a card or die value to be other than a whole number and strictly speaking the estimates should be rounded to the nearest unit. However, the answer here is left in decimal form so that the difference between estimates (*i*) and (*ii*) can be shown.

See Fig. A4.10 for the superimposition of the regression lines on the scattergraph in Fig. A4.8.

(*c*) In (*a*) estimates of card values for the two *extreme* die values were virtually the same, i.e. 5.70 and 6.39. Indeed, since decimal values are impossible in this case, rounding to the nearest unit does give identical estimates of 6. This only confirms a commonsense view of what is being said. If two variables are clearly unrelated (as must be the case when random card values and random die values are

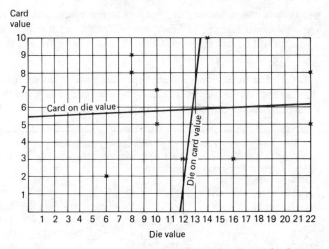

Figure A4.10 *Regression lines of data in Progress test 12, question 2*

recorded) then the value of one will be quite independent of the value of the other, and the best estimate of the former will be its mean value. Of course, such an estimate will be quite unreliable, but the error which results from using the mean will probably be smaller than that arising from the use of any other value. The best card value estimate, then, is the mean card value, i.e. 6.

This value of 6 is the best estimate of card value for all die values. This means that on a graph, having die values along the horizontal axis, the line of best fit is a *horizontal line* opposite 6 on the vertical axis. Considering now the estimation of *die values*, using the same arguments as above it is clear that the best estimate is the *mean die value*, no matter what the card value might be. On our graph this will give a *vertical line* opposite 12.8 on the horizontal axis. This means that the line of best fit suitable for estimating a *y* value from an *x* value is quite a different line from that suitable for estimating an *x* value from a *y* value.

In our example the regression lines are not quite horizontal and vertical. As the variables are quite unrelated they ought to be, but chance factors governing the values actually recorded resulted in a very slight associationship appearing to exist between card and die values. Such false slight associationship commonly arises in statistics due to these chance factors – an absolutely correct result is, in fact, as improbable as obtaining 500 'heads' and 500 'tails' in a 1,000 tosses of a coin (as will be seen in the next chapter, chance association of this kind is called *spurious correlation*).

3. (*a*) Here time, for once, is the dependent variable since forging time depends on the bar size. It is, therefore, designated as *y*, and the difference between the starting and finishing CSA is designated *x*.

Starting		Finishing		Difference	Time (mins)		
Size (ins)	CSA	Size (ins)	CSA	x	y	xy	x^2
9×4	36	4×1	4	32	16	512	1024
8×6	48	4×2	8	40	24	960	1600
4×4	16	1×1	1	15	9	135	225
8×2	16	2×2	4	12	12	144	144
8×8	64	4×4	16	48	29	1392	2304
7×7	49	5×2	10	39	21	819	1521
6×7	42	5×2	10	32	14	448	1024
				218	125	4410	7842

$$b = \frac{7 \times 4410 - 218 \times 125}{7 \times 7842 - 218^2} = 0.5$$

$$a = \frac{125 - 0.5 \times 218}{7} = 2.3 \text{ to one decimal place}$$

$$\therefore t = 2.3 + \frac{1}{2}(\text{CSA at start} - \text{CSA at finish})$$

(*b*) (*i*) Time to forge bar 2×3 from 6×6 = 2.3 + ½(6×6 − 2×3)
= 17.3 mins

Time to forge bar 2×3 from 5×5 = 2.3 + ½(5×5 − 2×3)
= 11.8 mins

∴ difference in time = 5.5 mins

(*ii*) Strictly speaking the *y on x* regression line should *not* be used to estimate *x* from *y*. Nevertheless, the question definitely asks for the regression line found above to be used, so:

Since $t = 2.3 + \frac{1}{2}(\text{CSA at start} - x)$ where $x = \text{CSA at finish}$,

then $18 = 2.3 + \frac{1}{2}(36 - x) = 2.3 + 18 - 0.5x = 20.3 - 0.5x$

$\therefore x = (20.3 - 18)/.5$

∴ CSA of finished bar = 4.6 sq. inches.

Progress test 14

1. First we compute all the figures needed for the formula:

$x^\star$	x^2	$y^\star$	y^2	xy
11	121	7	49	77
10	100	6	36	60
14	196	8	64	112
12	144	6	36	72
8	64	1	1	8
9	81	2	4	18
Σ 64	706	30	190	347

Now we can apply the formula:

$$r = \frac{n \times \Sigma xy - \Sigma x \times \Sigma y}{\sqrt{(n \times \Sigma x^2 - (\Sigma x)^2)(n \times \Sigma y^2 - (\Sigma y)^2)}}$$

$$= \frac{6 \times 347 - 64 \times 30}{\sqrt{(6 \times 706 - 64^2)(6 \times 190 - 30^2)}} = +0.88 †$$

Notes: *It is quite permissible to scale down or up either or both series of figures in order to simplify calculations. This scaling can be made by either dividing by a constant or by subtracting a constant or both. Observe that:

(*i*) the scaling of one series is completely independent of any scaling of the other series

(*ii*) since correlation is not concerned with absolute values – only degrees of change – no adjustment for such scaling needs to be made later in the calculations.

† *r* is not expressed in any units.

2.

x	x^2	y	y^2	xy
8	64	8	64	64
8	64	9	81	72
14	196	10	100	140
22	484	5	25	110
22	484	8	64	176
16	256	3	9	48
12	144	3	9	36

6	36	2	4	12
10	100	7	49	70
10	100	5	25	50
128	1928	60	430	778

$$r = \frac{10 \times 778 - 128 \times 60}{\sqrt{(10 \times 1928 - 128^2)(10 \times 430 - 60^2)}} = +.070$$

This type of correlation is *very low positive*, so low as to indicate that the variables are virtually uncorrelated (*see* **2(e)**).

Since there is no true correlation between the value of a thrown die and the value of a drawn card, such slight correlation that is given by the computations is obviously *spurious* (*see* **4**).

3.

Person	IQ ranking	Exam ranking
A	3	2
B	4	3 equal (i.e. 3.5*)
C	1	1
D	2	3 equal (i.e. 3.5*)
E	6	6
F	5	5

Spearman's coefficient of rank correlation $= 1 - \dfrac{6\Sigma d^2}{n(n^2-1)}$

Computation of Σd^2

IQ	Exam	d	d^2
3	2	1	1
4	3.5	0.5	0.25
1	1	0	0
2	3.5	1.5	2.25
6	6	0	0
5	5	0	0
			$\Sigma d^2 =$
			3.5

$\therefore$ Spearman's coefficient of rank correlation $= 1 - \dfrac{6 \times 3.5}{6 \times (6^2 - 1)} = +0.9$

In Question 1, r was $+0.88$. The difference arises through the fact that when ranking is used instead of the full set of figures there is some loss of information. This loss is reflected in the correlation values.†

Notes: *See **7**.

†The more accurate correlation value is, of course, the one computed from the full set of figures – in this case $+0.88$.

4. (a) In the layout below the written test rankings are coded as 'W', the interview rankings as 'I' and the job performance as 'J'.

				W v. J		I v. J	
Trainee	W	I	J	d	d²	d	d²
A	6	1	1	5	25	0	0
B	2	4	2	0	0	2	4
C	7	2	3	4	16	1	1
D	4	3	4	0	0	1	1
E	1	6	5	4	16	1	1
F	5	5	6	1	1	1	1
G	3	8	7	4	16	1	1
H	8	7	8	0	0	1	1
Total					74		10

Here $n = 8$

(i) Correlation between written test and job performance:

Spearman's coefficient of rank correlation

$$= 1 - \frac{6 \times 74}{8(8^2 - 1)} = 1 - \frac{444}{504} = \underline{+0.119}$$

(ii) Correlation between interview and job performance:

Spearman's coefficient of rank correlation

$$= 1 - \frac{6 \times 10}{8(8^2 - 1)} = \underline{+0.881}$$

(b) The correlation in (a)(i) shows that there is for all practical purposes no correlation between job performance and the written test – i.e. the test in no way predicts the person's later job performance. Conversely, the correlation in (a)(ii) shows that there is something of a positive correlation between job performance and the interview, although the small sample size (and the fact that

Spearman's coefficient of rank correlation tends to be significantly higher than the more conventional r) means that the predictive power of the interview should not be relied upon too heavily.

The decision that is suggested by the analysis is that the written test should be scrapped and engagements based on interview – though it would be wise to see if a better predictor couldn't be found in these circumstances.

One note, however, to the analysis. From the information given we know nothing of the people who were *not* engaged. If it had happened that, for instance, the trainee who had come out top in the written test had ranked bottom on performance, and vice versa, this on its own would not mean that the test was worthless but only that it failed to predict the performance of the *accepted* trainees. Assuming all eight trainees were of an acceptable standard, the written test may well have excluded 100 others who could not reach that standard – i.e. it could at least have separated out the acceptable from the unacceptable trainees. Strictly speaking, then, the analysis does not so much indicate that the test is necessarily worthless but only that it is inferior to the interview in predicting the job performance of those people who are actually accepted for training.

5.

x	y	x^2	y^2	xy
5	17	25	289	85
75	19	5625	361	1425
80	36	5650	650	1510

$$r = \frac{2 \times 1510 - 80 \times 36}{\sqrt{(2 \times 5650 - 80^2)(2 \times 650 - 36^2)}} = \frac{3020 - 2880}{\sqrt{4900 \times 4}} = \underline{\underline{+1}}$$

This answer is rather unexpected since the original figures do not, on the face of it, appear to be perfectly correlated. However, the reason for this answer is easily explained, for if there are only two points on a scattergraph the line of best fit is simply a line joining the points. Now in the case where all the points lie on the line of best fit there is perfect correlation. Thus, there is always perfect correlation where there are only two pairs of figures, as in this case.

The conclusion that follows from this is that the interpretation of r depends, among other things, upon the number of pairs of figures. Two pairs give perfect correlation, but each additional pair beyond this makes it increasingly difficult to find a line of best fit that falls close to all the points. Thus the value of r will tend to diminish as more pairs of figures are added. (Note, however, this change in the value of r is only of importance where the number of pairs of figures is small.)

6. No, an error need not have occurred. The following explanation could quite easily account for the result.

Hours of study is not the only factor influencing examination success. Maturity and intellectual ability could both be vital factors also. Now it is well known that people with high maturity and/or high intellectual ability need to study for fewer hours than their less fortunate colleagues in order to reach the same level of knowledge. If now the examination favours students with high maturity and/or intellectual ability then the success of such students will be greater than their colleagues, despite the extra hours of studying by the latter. Thus examination marks and hours of study would have a high negative correlation.

Note on answer: This example illustrates the difference between relationship and cause. Although fewer hours of study are related to examination success it is obviously quite wrong to argue that such success is *due* to the fewer hours' study, and that therefore the less one studies the better one's chances of success.

Progress test 15

1. (a) (i) Laspeyre price index:

Item	19-3 q_0	19-3 p_0(£)	19-3 p_0q_0(£)	19-4 p_1(£)	19-4 p_1q_0(£)	19-5 p_2(£)	19-5 p_2q_0(£)	19-6 p_3(£)	19-6 p_3q_0(£)
A	20	0.20	4.00	0.25	5.00	0.35	7.00	0.50	10.00
B	12	0.25	3.00	0.25	3.00	0.10	1.20	0.12½	1.50
C	3	1.00	3.00	2.00	6.00	2.00	6.00	2.00	6.00
Σ			10.00		14.00		14.20		17.50

Using the Laspeyre formula

Year	Laspeyre price index (19-3 = 100)
19-3	100
19-4	$\dfrac{14.00}{10.00} \times 100 = 140$
19-5	$\dfrac{14.20}{10.00} \times 100 = 142$
19-6	$\dfrac{17.50}{10.00} \times 100 = 175$

(ii) *Paasche price index:*

| | 19–3 | | 19–4 | | | 19–5 | | | | 19–6 | | |
Item	p_0 (£)	p_1 (£)	q_1	p_1q_1 (£)	p_0q_1 (£)	p_2 (£)	q_2	p_2q_2 (£)	p_0q_2 (£)	p_3 (£)	q_3	p_3q_3 (£)	p_0q_3 (£)
A	0.20	0.25	24	6.00	4.80	0.35	20	7.00	4.00	0.50	18	9.00	3.60
B	0.25	0.25	16	4.00	4.00	0.10	20	2.00	5.00	0.12½	16	2.00	4.00
C	1.00	2.00	2	4.00	2.00	2.00	3	6.00	3.00	2.00	4	8.00	4.00
Σ				14.00	10.80			15.00	12.00			19.00	11.60

Using the Paasche formula

Year	Paasche price index (19–3 = 100)
19–3	100
19–4	$\dfrac{14.00}{10.80} \times 100 = 130$
19–5	$\dfrac{15.00}{12.00} \times 100 = 125$
19–6	$\dfrac{19.00}{11.60} \times 100 = 164$

(iii) Weighted average of price relatives:

	Basic data 19-3			19-4		19-5			19-6	
Item	p_0 (£)	Weight	p_1 (£)	$\frac{p_1}{p_0} \times 100$ relative	Price relative × Weight	p_2 (£)	$\frac{p_2}{p_0} \times 100$ relative	Price relative × Weight	p_3 (£)	$\frac{p_3}{p_0} \times 100$ relative	Price relative × Weight
A	0.20	5	0.25	125	625	0.35	175	875	0.50	250	1,250
B	0.25	3	0.25	100	300	0.10	40	120	0.12½	50	150
C	1.00	2	2.00	200	400	2.00	200	400	2.00	200	400
Σ		10			1,325			1,395			1,800

Weighted average of price relatives (19–3 = 100)

Year	
19–3	100
19–4	$\frac{1,325}{10} = 132.5$
19–5	$\frac{1,395}{10} = 139.5$
19–6	$\frac{1,800}{10} = 180$

2. Revaluation of the cost of the plant calls for no more than the application of the formula given in **22**. The depreciation figures, however, cannot be dealt with in this manner (other than the 19–5 depreciation) since they are made up of amounts relating to plant bought in different years. Although one could analyse these depreciation figures into their constituent years and apply the revaluation formula to the analysed amounts, it is easier in this case to compute the new depreciation from scratch by computing for each year's plant purchases the total cumulative depreciation chargeable to the end of 19–9 on those purchases. This is done by using the formula:

Cumulative depreciation = Asset revaluation figure
$$\times \text{Number of years depreciation} \times 15\%$$

The revaluation and depreciation computation can then be set out in parallel as follows:

Year	Revaluation of plant bought in year	Depreciation chargeable on plant bought in year
19–5	$£35,000 \times \dfrac{268}{184} = £50,978$	$£50,978 \times 5 \times 15\% = £38,234$
19–6	$£42,000 \times \dfrac{268}{201} = £56,000$	$£56,000 \times 4 \times 15\% = £33,600$
19–7	$£88,000 \times \dfrac{268}{212} = £111,245$	$£111,245 \times 3 \times 15\% = £50,060$
19–8	$£51,000 \times \dfrac{268}{234} = £58,410$	$£58,410 \times 2 \times 15\% = £17,523$
19–9	$£82,000 \times \dfrac{268}{256} = £85,844$	$£85,844 \times 1 \times 15\% = £12,877$
	£362,477	£152,294

The answers to the question are, therefore:

(a) Revalued plant cost = £362,477
(b) Additional amount of depreciation to be charged
   = Total required depreciation on revaluated assets
     − existing total charged
   = £152,294 − £118,650 = £33,644

3. The following could explain why the company's sales have fallen relative to TV licences:

(*a*) *A fall in the market share.* If TV licences reflect an increase in the sales of television sets, then the relative drop in the company's sales could be due to the fact that its competitors are selling relatively more sets themselves.

(*b*) *Earlier sales were replacement sales.* To appreciate the significance of this point, imagine that the only sales in the country were sets that replaced existing sets. In this case no first-time licences would be taken out – and so the television licence index would remain unchanged whereas the sales index would increase if replacement sales exceeded previous years' replacement sales – and fall if otherwise. So a static licence index would not mean no sales of television sets were being made, but that on the whole only replacement sets were being sold – and in that case a company's sales performance would depend on replacement sets.

(*c*) *More secondhand sets being bought by first-time users.* If in this case the sellers of the sets have a second set, they will not buy a new licence. So licences can increase without any *new* sets being sold at all.

(*d*) *Improved detection rate of people without licences.* In this situation not only will such people take out television licences without buying sets but many people who fear they may now be caught as a result of this improvement will do the same thing. So licences can increase without *any* sets being traded – new or replacement.

(*e*) *More expensive sets are being sold by the company.* Clearly, the more expensive sets will sell in fewer numbers, though licences may still increase considerably as a result of another company's sales of cheaper sets. However, the drop in sales numbers may be accompanied by an increase in total sales revenue so that despite the implication that sales fall is bad, in fact the company may be doing very much better.

4. (*a*)(*i*)

Period ending:	31/12/76	30/6/77	31/12/77	30/6/78	31/12/78	30/6/79	31/12/79
Index	100	105	114	123	130	139	149
Increase		5	9	9	7	9	10
Percentage		5:100	9:105	9:114	7:123	9:130	10:139
increase		=5%	=8.6%	=7.9%	=5.7%	=6.9%	=7.2%

So the six-month period with the greatest percentage increase is the period ending 31/12/77 – which has 8.6%.

(*ii*)

Index at 31/12/76	100
14% increase	14
Index at 31/12/77	114
14% increase	16
Index at 31/12/78	130
14% increase	18.2
Index at 31/12/79	148.2

As can be seen, all three figures computed on the basis of a 14% annual increase are consistent with the index figures given. Only the last is a little below the given index which indicates that the actual rate is fractionally above 14%.

If this rate were maintained then by 31/12/80 the index would be 149 + 14% of 149 = 149 + 20.9 = say, 170, and by 31/12/81 it would be 170 + 14% of 170 = 170 + 23.8 = say, 194. (The roundings allow for the fact that the true rate is a little above 14%.) Note: This part of the question can also be answered by using the equation given in (*b*).

(*b*)

Cost	Content of Total cost	31/12/76 Index	31/12/76 Var. cost	31/12/79 Index	31/12/79 Var. cost
Materials	40%	100	40	149	59.6
Wages	40%	100	40	163	65.2
Var. O'h'd	20%	100	20	127	25.4
			100		150.2

∴ Average annual ratio change in the variable cost (*see* **24**):

$$\sqrt[n]{\text{total ratio change}} = \sqrt[3]{150.2/100} = 1.145$$

∴ Average annual percentage increase
= Percentage increase from 1 to 1.145
= 14.5%

Progress test 17

1. (a) Since both series cover the same time period, space can be saved by laying out both series in the same table:

Year	$d\star$	d^2	A	Ad	B	Bd
1978	−4	16	25	−100	200	−800
1979	−3	9	31	−93	196	−588
1980	−2	4	36	−72	194	−388
1981	−1	1	44	−44	190	−190
1982	0	0	48	0	189	0
1983	+1	1	53	+53	185	+185
1984	+2	4	60	+120	184	+368
1985	+3	9	62	+186	181	+543
1986	+4	16	67	+268	180	+720
$\Sigma =$		60	426	+318	1,699	−150

Apply formulae for $y = a + bd$ equation.

A series:

$$a = \overline{A} = \frac{\Sigma A}{n} = \frac{426}{9} = \underline{47.3} \qquad b = \frac{\Sigma Ad}{\Sigma d^2} = \frac{+318}{60} = \underline{+5.3}$$

$\therefore$ Population A (000s) $= \underline{47.3 + 5.3d}$

B series:

$$a = \overline{B} = \frac{\Sigma B}{n} = \frac{1,699}{9} = \underline{188.8} \qquad b = \frac{\Sigma Bd}{\Sigma d^2} = \frac{-150}{60} = \underline{-2.5}$$

$\therefore$ Population B (000s) $= \underline{188.8 - 2.5d}$†

(b) When the populations are equal, then population A = population B, i.e.

$$47.3 + 5.3d = 188.8 - 2.5d$$
$$\therefore 5.3d + 2.5d = 188.8 - 47.3$$
$$\therefore d \simeq 18$$

$\therefore$ Population will be equal in $1982 + 18 = 2000$

In that year the populations of A and B are both estimated to be $47.3 + 5.3d^{\ddagger} = 188.8 - 45 \simeq \underline{143,000}$

Notes: *Mid-point of series = mid-point of 1982.

†Note that the minus sign indicates a *downward* trend.

‡Since the populations are equal, substitution in only one equation is needed.

2. (a) (i) Seasonal variations

Year	Quarter	Demand (tonnes)	MAT	Moving average	Centred average*	Seasonal variation Multiplicative	Additive
19–4	1	218					
	2	325					
			1,064	266			
	3	273			294	93	−21
			1,290	322½			
	4	248			355	70	−107
			1,550	387½			
19–5	1	444			409	108½	+35
			1,722	430½			
	2	585			448	131	+137
			1,859	464¾			
	3	445			492	90½	−47
			2,075	518¾			
	4	385			552	70	−167
			2,342	585½			
19–6	1	660			608	109	+52
			2,520	630			
	2	852			648	131½	+204
			2,660	665			
	3	623					
	4	525					

Multiplicative averaging

Quarter	1	2	3	4	Total
19–4	–	–	93	70	
19–5	108½	131	90½	70	
19–6	109	131½	–	–	
Total	217½	262½	183½	140	
Average	108¾	131¼	91¾	70	401¾
Adjusted	108	131	91	70	400

Additive averaging:

Quarter	1	2	3	4	Total
19–4	–	–	– 21	– 107	
19–5	+ 35	+ 137	– 47	– 167	
19–6	+ 52	+ 204	–	–	
Total	+ 87	+ 341	– 68	– 274	
Average	+ 43½	+ 170½	– 34	– 137	43
Adjusted	+ 40	+ 150	– 40	– 150	0

The seasonal variations are, therefore:

	Qr 1	Qr 2	Qr 3	Qr 4
Multiplicative	108	131	91	70
Additive	+ 40	+ 150	– 40	– 150

(*ii*) *The trend*‡

$$y = a + bd$$

Year	Demand (y§)	d	yd	d^2
19–4	1,064	– 1	– 1,064	1
19–5	1,859	0	0	0
19–6	2,660	+ 1	+ 2,660	1
	$\Sigma y = 5,583$		$\Sigma yd = + 1,596$	$\Sigma d^2 = 2$

$a = y = 5,583/3 = \underline{1,861}$

and $b = \Sigma yd / \Sigma d^2 = 1,596/2 = 798$

$\therefore y = \underline{1,861 + 798d}$

(*b*) (*i*) *Total demand for* 19–8

19–8 is 3 years' deviation from 19–5

$\therefore$ Demand $y = 1,861 + 798 \times 3$

$= 4,255$, (say) $\underline{4,250 \text{ tonnes}}$★★

(*ii*) *Demand for last quarter* (4) *of* 19–8:

First adjust the trend formula for quarters, i.e.:

$y' = 1,861/4 + 798/4^2 \times d'$ (where $d' = $ Deviation in quarters)

$\therefore y' = 465 + 50d'$ (rounded).

Midpoint of series = Midpoint of 19–5, i.e. end of Quarter 2, 19–5.

$\therefore$ Deviation of Quarter 4, 19–8, from the midpoint of the series:

	Quarters
From end of Quarter 2, 19–5, to the end of 19–5	2
Quarters of the years 19–6 and 19–7	8
From beginning of 19–8 to end of Quarter 3	3
End of Quarter 3 to midpoint of Quarter 4	½
	+ 13½

$\therefore$ The trend for Quarter 4, 19–8:

$y' = 465 + 50 \times 13\frac{1}{2} = \underline{1{,}140}$

$\therefore$ Demand for last quarter of 19–8:

= Trend × Seasonal variation

= $1{,}140 \times 70/100 = 798$, (say) <u>800 tonnes</u>

Notes: *These figures are rounded.

‡In practice, 3 years is too short a duration to obtain a reliable trend equation.

§These figures are the sum of the four quarters for each year.

**Since the estimated figures cannot be accurate to a tonne, it is better to round to a reasonable number.

If we wish to allow for the random variations then these must first be computed:

Year	Q	Actual	d	Theoretical*	Random variation (%)
19–4	1	218	$-5\frac{1}{2}$	205	$+6\frac{1}{2}$
	2	325	$-4\frac{1}{2}$	314	$+3\frac{1}{2}$
	3	273	$-3\frac{1}{2}$	263	$+4$
	4	248	$-2\frac{1}{2}$	238	$+4$
19–5	1	444	$-1\frac{1}{2}$	421	$+5$
	2	585	$-\frac{1}{2}$	576	$+2$
	3	445	$+\frac{1}{2}$	446	0
	4	385	$+1\frac{1}{2}$	378	$+2$
19–6	1	660	$+2\frac{1}{2}$	637	$+4$
	2	852	$+3\frac{1}{2}$	838	$+2$
	3	623	$+4\frac{1}{2}$	628	-1
	4	525	$+5\frac{1}{2}$	518	$+1$

*Theoretical demand $=(465+50d')\times$ seasonal variation.

In our estimate, then, we should allow a range of, say, $+5\%$, i.e. estimate $=800-840$ tonnes.

3. Formula for b in the regression line equation:

$$b=\frac{n\times\Sigma yx-\Sigma y\times\Sigma x}{n\times\Sigma x^2-(\Sigma x)^2}$$

Let the x series of values measure the deviations from the midpoint of the x series.

Now, since in a time series the deviations are all in equal steps ranged symmetrically around the midpoint of the series (e.g. $-2\frac{1}{2}$, $-1\frac{1}{2}$, $-\frac{1}{2}$, $\frac{1}{2}$, $1\frac{1}{2}$, $2\frac{1}{2}$) then their sum must be zero, i.e. $\Sigma x=0$

So $b=\dfrac{n\times\Sigma yx-\Sigma y\times 0}{n\times\Sigma x^2-(0)^2}$

(note that Σx^2 is *not* zero since squaring the negative deviations gives positive values)

$$=\frac{n\times\Sigma yx}{n\times\Sigma x^2}=\frac{\Sigma yx}{\Sigma x^2}$$

Now write xy in place of yx and d in place of x

$$\therefore\ b=\Sigma dy/\Sigma d^2$$

Next, the formula to find a is:

$$a=\frac{\Sigma y-b\times\Sigma x}{n}=\frac{\Sigma y-b\times 0}{n}=\Sigma y/n=\bar{y}$$

But these are the formulae used to compute a trend line.

$\therefore$ The trend line is the regression line of y *on* x, where x measures the deviations of the periods from the midpoint of the series.

4. $\underline{8}, \times0.75+12\times0.25=\underline{9}, \times0.75+15\times0.25$.

$=\underline{10.5}, \times0.75+14\times0.25$

$=\underline{11.38}, \times0.75+17\times0.25=\underline{12.78}, \times0.75+20\times0.25$

$=\underline{14.58}, \times0.75+25\times0.25=\underline{17.18}, \times0.75+23\times0.25$

$=\underline{18.64}, \times0.75+28\times0.25=\underline{20.98}, \times0.75+30\times0.25$

$=\underline{23.24}, \times0.75+33\times0.25=\underline{25.68}$

Average lag $= \dfrac{1-0.25}{0.25} = 3$ periods.

Progress test 18

2. The probabilities can only be relative frequencies.

The table shows that there are 958 males of 25 years. So:

(a) Since 905 males reach the age of 45, the probability that a given male will attain 45 is 905/958 = <u>0.945</u>

(b) Since there are 905 males at 45 but only 413 at 75, then 905 − 413 = 492 of those males do not attain 75. So of the 958 alive at 25, 492 will die between 45 and 75. So probability of one of these 25-year-olds dying between these ages is 492/958 = <u>0.514</u>

(c) Of the 958 alive at 25 only 680 are alive at 65, i.e. 958 − 680 = 278 die. So probability of a 25-year old not attaining 65 years is 278/958 = <u>0.290</u>

3. Using the method of relative frequency and the fact that there is a total of 2,200 members from which the random selection is made:

(a) Number skilled and in favour = 800. So probability = 800/2200 = <u>0.364</u>

(b) Number undecided = 500. So probability = 500/2200 = <u>0.227</u>

(c) Number unskilled = 900. Number opposed = 800. Total = 1700. However, this includes 600 who are both unskilled and opposed. So total either unskilled or opposed = 1700 − 600 = 1100. So probability = 1100/2200 = <u>0.500</u>

4. (a) (i) $\frac{1}{2} \times \frac{1}{2} = \underline{\frac{1}{4}}$

   (ii) $1/6 \times 1/6 = \underline{1/36}$

(b) $1/6 + 1/6 = \underline{1/3}$

(c) The probability of drawing the ace of spades or clubs or hearts or diamonds = $1/52 + 1/52 + 1/52 + 1/52 = 1/13$

(d) Sequential events are HT, TH
∴ Probability = $(\frac{1}{2} \times \frac{1}{2}) + (\frac{1}{2} \times \frac{1}{2}) = \underline{\frac{1}{2}}$

(e) Sequential events giving a total of 11: 5, 6; 6, 5
∴ Probability, therefore, of 11 = $(1/6 \times 1/6) + (1/6 \times 1/6) = 1/18\star$

(*f*) Sequential events giving total of 7: 1, 6; 2, 5; 3, 4; 4, 3; 5, 2; 6, 1
∴ Probability of 7
$$=(1/6 \times 1/6) + (1/6 \times 1/6) + (1/6 \times 1/6) + (1/6 \times 1/6) + (1/6 \times 1/6) + (1/6 \times 1/6) = \underline{1/6}$$

Note: *It should be appreciated that the probability of throwing a 5 and a 6 is *not* the same as throwing two 6s. The reason is that there are two different sequential events giving 5 and 6 (5, 6 and 6, 5) but only one of a double 6 (6, 6).

5. (*a*) Probability both will be alive $= 5/8 \times 5/6 = \underline{25/48}$

(*b*) Probability at least one of them will be alive $= 1 -$ probability both be dead $= 1 - (3/8 \times 1/6) = \underline{15/16}$

(*c*) Probability only wife will be alive $=$ Probability husband dead *and* wife alive $= 3/8 \times 5/6 = \underline{5/16}$

6. Probability that A will obtain a given contract $= 10/20 = 0.5$, that B will obtain a given contract $= 6/20 = 0.3$ and that neither will obtain a given contract $= 4/20 = 0.2$.

(*a*) Probability that A will obtain all three contracts
$$= 0.5 \times 0.5 \times 0.5 = \underline{0.125}$$

(*b*) Probability that B will obtain at least one contract $= 1 -$ probability that B obtains no contracts
$$= 1 - (0.7 \times 0.7 \times 0.7) = \underline{0.657}$$

(*c*) Probability that 1st and 2nd contract, or 1st and 3rd contract, or 2nd and 3rd contract obtained by neither*
$$= (0.2 \times 0.2 \times 0.8) + (0.2 \times 0.8 \times 0.2) + (0.8 \times 0.2 \times 0.2) = \underline{0.096}$$

(*d*) Probability that A obtains 1st contract, B 2nd contract and A 3rd contract $= 0.5 \times 0.3 \times 0.5 = \underline{0.075}$

Note: *This assumes that the remaining contract *is* to A or B.

7. (*a*) P(A and B but not C) $= 0.3 \times 0.4 \times (1 - 0.6) = \underline{0.048}$

(*b*) P(A and B regardless of C) $= 0.3 \times 0.4 \times 1$ (since it is certain that the company will either win or not win C) $= \underline{0.120}$

(*c*) P(C and D) $=$ P($\bar{\text{A}}$ and $\bar{\text{B}}$ and C and D) $= 0.7 \times 0.6 \times 0.6 \times 0.8 = \underline{0.2016}$

(*d*) P(C) $=$ P($\bar{\text{A}}$ and $\bar{\text{B}}$ and C and $\bar{\text{D}}$), since in these circumstances the company would have bid for D $= 0.7 \times 0.6 \times 0.6 \times 0.2 = \underline{0.0504}$

(*e*) $P(D) = P(\overline{A} \text{ and } \overline{B} \text{ and } \overline{C} \text{ and } D) = 0.7 \times 0.6 \times 0.4 \times 0.8 = \underline{0.1344}$

8. $P(\overline{A} \text{ and } \overline{B}) = 0.1 \times 0.1 = 0.01$. So $P(A \text{ or } B \text{ or both})$
$= 1 - P(\overline{A} \text{ and } \overline{B}) = 1 - 0.01$
$= 0.99$
Similarly, $P(C \text{ or } D \text{ or both}) = 0.99$
So probability $P(A \text{ or } B \text{ or both})$ *and* $P(C \text{ or } D \text{ or both})$
$= 0.99 \times 0.99$
$= 0.9801$

9. If we let Y = 'Yes' and N = 'No' to the question 'does the retailer allow himself to be interviewed?', then all the possible sequential events are listed in the left-hand column below. Note that once the survey worker has achieved his quota he stops visiting retailers and the probability of subsequent potential interviewees accepting or refusing does not arise.

		Visits					
Quota	*Question*	1 2 3 4			*Probability*		
Filled	(*a*)(*i*)	Y Y – –			0.6×0.6		$= 0.3600$
	(*a*)(*ii*)	N Y Y –			$0.4 \times 0.6 \times 0.6$	$= 0.1440$	$\Big\} = 0.2880$
		Y N Y –			$0.6 \times 0.4 \times 0.6$	$= 0.1440$	
	(*a*)(*iii*)	N N Y Y			$0.4 \times 0.4 \times 0.6 \times 0.6$	$= 0.0576$	
		N Y N Y			$0.4 \times 0.6 \times 0.4 \times 0.6$	$= 0.0576$	$\Big\} = 0.1728$
		Y N N Y			$0.6 \times 0.4 \times 0.4 \times 0.6$	$= 0.0576$	
Not filled	(*b*)	N N Y N			$0.4 \times 0.4 \times 0.6 \times 0.4$	$= 0.0384$	
		N Y N N			$0.4 \times 0.6 \times 0.4 \times 0.4$	$= 0.0384$	$\Big\} = 0.1792$
		Y N N N			$0.6 \times 0.4 \times 0.4 \times 0.4$	$= 0.0384$	
		N N N –			$0.4 \times 0.4 \times 0.4$	$= 0.0640$	

Cross check: Probability quota filled *or* not filled   1.0000
(Alternatively answer to (*b*) can be found as $1 - P$(quota filled)
$$= 1 - (0.36 + 0.288 + 0.1728)$$

10. *Probabilities*:
Delivery: In 1 week = 0.2. In 2 weeks = 0.5. In 3 weeks = 0.3
Usage: 1 unit in week = 0.6. 2 units in week = 0.4

(*a*) (*i*) Since delivery times are mutually exclusive events then:

$P(2 \text{ weeks or more}) = P(2 \text{ weeks} \cup 3 \text{ weeks}) = P(2 \text{ weeks}) + P(3 \text{ weeks})$
$$= 0.5 + 0.3 = \underline{0.8}$$

(*ii*) When deliveries are *not* made in week 1 then on 5 occasions delivery will be in week 2 and on 3 occasions in week 3 (these are the 0.5 and 0.3 probabilities multiplied by 10 so as to give us whole numbers). So out of 8 occasions 5 deliveries will be in week 2. So, using method of relative frequency, probability of a delivery in week $2 = 5/8 = \underline{0.625}$

(*iii*) (1) The compound event '3 items in 2 weeks' can be analysed into the following sequential events and probabilities:

Sequential events		Probability
(1 item week 1) $\cap$ (2 items week 2)		$0.6 \times 0.4 = 0.24$
(2 items week 1) $\cap$ (1 item week 2)		$0.4 \times 0.6 = \underline{0.24}$
$\therefore$ P(3 items in 2 weeks)	=	$\underline{\underline{0.48}}$

(2) The compound event '4 items in 2 weeks' can only be analysed into the single sequential events (2 items week 1) *and* (2 items week 2) which have a probability of $0.4 \times 0.4 = \underline{0.16}$

(*b*) The company will avoid running out of stock if either:

(*i*) it receives a delivery after 1 week regardless of usage in that week; *or*

(*ii*) it uses 1 unit in week 1 *and* 1 unit in week 2 *and* receives a delivery after two weeks.

Now the deliveries are mutually exclusive so these events are mutually exclusive. So the sequential events and their probabilities are:

Sequential events	Probability
(Delivery in 1 week)	0.20
(1 unit week 1)$\cap$(1 unit week 2)$\cap$(Delivery week 2)	$0.6 \times 0.6 \times 0.5 = \underline{0.18}$
$\therefore$ Probability company avoids running out of stock	0.38
So probability company *will* run out of stock $= 1 - 0.38 = \underline{0.62}$	

Progress test 19

1.

Black marble drawn	Probability		Winnings	Expectation (£)
1st draw	1/5	= 1/5	£0	0
2nd draw	4/5 × 1/4	= 1/5	£1	0.2
3rd draw	4/5 × 3/4 × 1/3	= 1/5	£2	0.4
4th draw	4/5 × 3/4 × 2/3 × 1/2	= 1/5	£3	0.6
5th draw	4/5 × 3/4 × 2/3 × 1/2 × 1	= 1/5	£4	0.8
		1		2.0

So the player has an expectation of £2

Note: We could have obtained the probability figures more easily by appreciating that any drawing of all the marbles would give us one of five equiprobable sequences of 5 marbles (since the black marble is no different from a white marble as far as the selection goes). So the five positions the black marble can take are all equiprobable – and so the probability of a given position is 1/5.

2. A priori we can say that on average the players will bet equally on all six symbols.

Assume that at each trial there is a stake of £1 on each symbol. The banker, then, receives £6. His payments and their probabilities are as follows:

(a) *3 faces all different symbols.* Here he will pay to 3 players their stake plus as much again, i.e. £6. So he nets £0. And the probability is:

Probability 2nd die not same as 1st × probability 3rd die not same as 1st and 2nd = 5/6 × 4/6 = 20/36.

(b) *3 faces all the same symbol.* Here he will pay 1 player his stake plus 3 times as much, i.e. £4. So he wins £2 net. And this probability is:

Probability 2nd die same as 1st × probability 3rd die same as 1st = 1/6 × 1/6 = 1/36.

(c) *2 faces same symbol and 3rd different.* Here he will pay 1 player his stake plus twice as much and another his stake plus as much, i.e. £3 + £2 = £5. And this probability is as follows:

(*i*) 2nd face same as 1st, 3rd different	$= 1/6 \times 5/6 =$ 5/36
(*ii*) 2nd face different, 3rd face same as 1st	$= 5/6 \times 1/6 =$ 5/36
(*iii*) 2nd face different, 3rd face same as 2nd	$= 5/6 \times 1/6 =$ $\underline{\text{5/36}}$
	15/36

So the banker's expectation can be found as follows:

Event	Probability	Net gain (£)	Expectation (£)
3 faces all different	20/36*	0	0
3 faces all the same	1/36	2	2/36
2 faces only the same	$\underline{\text{15/36}}$	1	$\underline{\text{15/36}}$
	1		17/36

So the banker's expectation on a game with a £6 stake = £17/36 = £0.472. And this expressed as expectation per £1 staked = 0.472/6 = £0.079, or a return of nearly 8 per cent of the total money staked.

Note: *Strictly speaking, we do not need to compute this probability for since the net gain is £0 the expectation will be £0 whatever the probability is. However, finding it does enable us to cross-check that the sum of our probabilities is 1.

3. The marble chosen from A must be either black or white. We have, then, two sequential events – one starting with a white marble and one with a black marble. The analysis, then, runs as follows:

Marble from A	Probability	Marbles now in B	P(White marble)	Event	Probability
White	⅓	1B + 3W	¾	W ∩ W	$\frac{1}{3} \times \frac{3}{4} = \frac{1}{4}$
Black	⅔	2B + 2W	½	B ∩ W	$\frac{2}{3} \times \frac{1}{2} = \frac{1}{3}$
Probability final marble is white					7/12

If the probability of the final marble being white is 7/12 then the probability of the final marble being black is $1 - 7/12 = 5/12$.

So expectation $= (7/12 \times 60\text{p}) + (5/12 \times -60\text{p}) = \underline{10\text{p}}$

4. (*a*) The only way 8p can be built up with 3 coins is by drawing a 5p coin, a 2p coin and a 1p coin. The sequential events and conditional probabilities, then, are as follows:

Sequential event			Probability	
5p	2p	1p	$4/20 \times 6/19 \times 10/18$	
5p	1p	2p	$4/20 \times 10/19 \times 6/18$	
2p	5p	1p	$6/20 \times 4/19 \times 10/18$	$=6 \times \dfrac{240}{20 \times 19 \times 18} = 0.211$
2p	1p	5p	$6/20 \times 10/19 \times 4/18$	
1p	5p	2p	$10/20 \times 4/19 \times 6/18$	
1p	2p	5p	$10/20 \times 6/19 \times 4/18$	

$\therefore$ probability of drawing exactly 8 p with 3 coins = $\underline{0.211}$

(*b*) If none of the three coins is a 5p piece then the maximum total value of all three cannot exceed 6p. Conversely, if only one of the coins is 5p then the very minimum total value of all three is 7p. So probability of total value being less than 7p is probability of no 5p pieces among the three coins, i.e. $16/20 \times 15/19 \times 14/18 = \underline{0.491}$.

5. (*a*) Note that the probability of my being late is the probability of my being less than 5 minutes late + the probability of my being 5 or more minutes late (since these are alternative exclusive events) = $0.2 + 0.7 = 0.9$. Since in this question the probability of arriving is 1 (certain), we need to know what the missing 0.1 is. Clearly, here it is the probability of not being late – when, of course, no reprimand will be given. So:

Sequential event	Probability
Early, no reprimand	$0.1 \times 1 = 0.10$
Less than 5 mins, no reprimand	$0.7 \times 0.6 = 0.42$
5 mins or more, no reprimand	$0.2 \times 0.1 = 0.02$

$\therefore$ Probability of no reprimand = $\underline{0.54}$

(Note alternative calculation: Probability of no reprimand = 1 – probability of reprimand (as computed below).)

(*b*) Probability of reprimand:

Sequential event	Probability
Less than 5 mins, reprimand	$0.7 \times 0.4 = 0.28$
5 mins or more, reprimand	$0.2 \times 0.9 = 0.18$

$\therefore$ Probability of reprimand = $\underline{0.46}\star$

∴Probability of being less than 5 minutes late *given*

$$\text{reprimand} = P(\text{Less than five mins late}|\text{reprimand}) = \frac{0.28}{0.46}$$

$$= \underline{\underline{0.609}}$$

Note: *Strictly speaking we should account for the third possible sequential event – the 'early, reprimand' event. However, since this comes to $0.1 \times 0 = 0$, it can be ignored in the layout.

6. First we must find the probability of an unreasonable bill. So we have:

Sequential event	Probability
Car unused by son, bill unreasonable	$0.7 \times 0.8 = 0.56$
Car used by son, bill unreasonable	$0.3 \times 1.0 = 0.30$

∴ Probability of unreasonable bill $= \underline{0.86}$

The best way to proceed now is to find the probability that my son did *not* use the car *given an unreasonable bill*. So:

$$P(\text{car unused by son}|\text{unreasonable bill}) = \frac{0.56}{0.86} = 0.651$$

Since the question is, what is the probability that my son used the car at all during the three months, we can say this is $1 -$ probability he never used it. The probability that he never used it in any of the three months, given that the bill was unreasonable in each month, is the probability that he never used it in the 1st month *and* never used it in the 2nd month *and* never used it in the third month $= 0.651 \times 0.651 \times 0.651 = 0.276$.

∴ Probability my son used the car to a greater or lesser extent
$$= 1 - 0.276 = \underline{0.724}$$

7. First note that since the three machines collectively can produce 600 units per hour, of which 300 are from A, 200 from B and 100 from C, the probabilities of a unit being allocated to these machines is $\frac{300}{600} = \frac{1}{2}$, $\frac{200}{600} = \frac{1}{3}$ and $\frac{100}{600} = \frac{1}{6}$ respectively.

(*a*) Probability a unit produced on A *and* emerges substandard
$= P(\text{unit allocated to A}) \times P(\text{unit emerges substandard}|\text{unit allocated to A})$

$$= \frac{1}{2} \times \frac{1}{10} = \underline{0.05}.$$

(*b*) Probability a unit emerges substandard = Probability made on A *and* is substandard, *or* made on B *and* is substandard *or* made on C *and* is substandard = P(allocated to A) × P(substandard | A) + P(allocated to B) × P(substandard | B) + P(allocated to C) × P(substandard | C)

$$= [\tfrac{1}{2} \times \tfrac{1}{10}] + [\tfrac{1}{3} \times \tfrac{15}{100}] + [\tfrac{1}{6} \times \tfrac{24}{100}]$$
$$= 0.05 + 0.05 + 0.04 = \underline{0.14}.$$

(*c*) This is a straight application of Bayes' Theorem. Since the probability of a given unit being allocated to A and proving substandard is 0.05 (*see* (*a*)) and the probability that a unit will in any event be substandard is 0.14 (*see* (*b*)) the probability a substandard unit was in fact produced on A – i.e. P(unit allocated to A | unit substandard) – is:

$$\frac{0.05}{0.14} = \underline{0.357}.$$

Note: In other words the probability is greater than ⅓ despite the fact that A has the lowest rate of substandard units. The reason the probability is so high is, of course, because A produces by far the most units and although its substandard *rate* is low, the proportion of substandard units it actually produces relative to the substandard units from the other machines is high.

8. (*a*) Let P1, P2 and P3 symbolise acceptance of a perfect bottle at points 1, 2 and 3, and F1, F2 and F3 the acceptance of a faulty bottle at these three points.

Now if there is a 0.02 probability that a perfect bottle will be rejected there is only a 0.98 probability that it will be accepted. And, of course, there is a 0.02 probability that a faulty bottle will be accepted. So P(P1), P(P2) and P(P3) are all 0.98 and P(F1), P(F2) and P(F3) are all 0.02.

(*i*) Probability that a perfect bottle will be passed through all points =

P(P1) *and* P(P2) *and* P(P3) = 0.98 × 0.98 × 0.98 = $\underline{0.9412}$

(*ii*) Probability that a bottle faulty in colour and with a crack will be passed =

P(P1) *and* P(F2) *and* P(F3) = 0.98 × 0.02 × 0.02 = $\underline{0.0004}$

(*iii*) Probability that a bottle faulty in size will be passed =

P(F1) *and* P(P2) *and* P(P3) = 0.02 × 0.98 × 0.98 = <u>0.0192</u>

(*b*) Probability unit produced on old machine, P(Old) = 0.35
Probability unit produced on new machine, P(New) = 0.65
And P(Defective | Old) = 0.08, while P(Defective | New) = 0.02

∴ P(Old | Defective)

$$= \frac{P(Old) \times P(Defective \mid Old)}{P(Old) \times P(Defective \mid Old) + P(New) \times P(Defective \mid New)}$$

$$= \frac{0.35 \times 0.08}{0.35 \times 0.08 + 0.65 \times 0.02} = 0.683$$

So probability that the defective unit was produced on the older machine = <u>0.683</u>

9. This is solvable using Bayes' Theorem.

Probabilities of prior events are: P(Defect) = 0.02, P(No defect) = 0.98.

Probabilities of the subsequent event 'Fail' are:
P(Fail | Defect) = 1.
P(Fail | No defect) = 0.05.

If the subsequent event is 'Fail' then the probability of having the defect is:

P(Defect | Fail)

$$= \frac{P(Defect) \times P(Fail \mid Defect)}{P(Defect) \times P(Fail \mid Defect) + P(No\ defect) \times P(Fail \mid No\ defect)}$$

$$= \frac{0.02 \times 1}{0.02 \times 1 + 0.98 \times 0.05} = \frac{0.02}{0.069} = \underline{0.29}$$

So the individual is less likely to have the defect if he fails the test than have it.

If this seems puzzling, think of it like this. If there is a population of 10,000 then 2 per cent = 200 people have the defect and all these will fail the test. Of the remaining 9,800 5 per cent = 490 will also fail the test. So a total of 200 + 490 = 690 will fail but since only 200 will have the defect the probability that someone who fails will have the defect is only 200/690 = 0.29.

This illustrates the common situation where, even if only a small proportion of healthy people are liable to fail a diagnostic test, the probability that a person failing the test is healthy is, in fact, quite high.

10. (*a*) (*i*) Independent. The result of one draw is in no way affected by a previous draw – neither does it affect a subsequent draw.

(*ii*) Dependent. Tax paid depends on salary in so far as one's tax is based upon one's salary.

(*iii*) Dependent. Experience shows that one is more liable to have an accident when drunk than when sober.

(*iv*) Independent. There is no connection between having large feet and being an accountant. (Note that there may, however, be a connection between large feet and membership of a more physically demanding profession.)

(*v*) Dependent. Two mutually exclusive events are dependent inasmuch as if one occurs then the other cannot.

(*b*) Probability of a component *not* being defective $= 1 - 0.01 = 0.99$

$\therefore$ Probability of all 4 components not being defective at the same time $= 0.99^4 = 0.9606$

$\therefore$ Probability that this is not so and hence the toy defective $= 1 - 0.9606$

$= \underline{0.0394}$

(*c*) Inspector inspects in the ratio of 3 from C, 1 from B and 1 from A. So:

(*i*) $P(A) = 1/5 = \underline{0.2}$

(*ii*) P(A) and P(Defective | A) $= 0.2 \times 0.1 = 0.02$
P(B) and P(Defective | B) $= 0.2 \times 0.1 = 0.02$
P(C) and P(Defective | C) $= 0.6 \times 0.2 = \underline{0.12}$

$\qquad$ P(Tube defective) $= \qquad \underline{0.16}$

(*iii*) This calls for the application of Bayes' Theorem, and the formula in **14**. Note that the lower part of the formula fraction – the probability of there being a defective at all – has been computed in (*ii*). So:

$$P(A \mid \text{Defective}) = \frac{P(A) \times P(\text{Defective} \mid A)}{P(\text{Defective})} = \frac{0.2 \times 0.1}{0.16} = \underline{\underline{0.125}}$$

11. (*a*) The coin is more likely to fall 'tails' since the law of averages asserts that in the long run the proportions of 'heads' and

'tails' will both be 0.5. So henceforth the number of 'tails' will exceed the number of 'heads' and so a 'tail' is more likely than a 'head' on the next toss.

(*b*) The coin is equally likely to fall a 'tail' as a 'head' since a coin has no memory of its past history and the next toss is as likely as the first to be a 'head' or a 'tail'. The a priori probability of a coin falling 'tails' is 0.5.

(*c*) The coin is more likely to fall a 'head' than a 'tail since the probability of an event can be measured by the past relative frequency of the event to the total number of trials. As the coin has fallen 'heads' 10 times out of 10 trials, the relative frequency is $10/10 = 1$, and while the relative frequency will be a very rough and ready value in such a short sequence of trials, nevertheless its current value of 1 for a 'head' gives us grounds for believing that on the next toss a 'tail' is less likely to fall than a 'head'.

The first argument is totally fallacious (*see* **1**). It is, though, difficult to decide between the next two. The question really is, is the coin a fair one? The chance of a fair coin falling the same way as its first fall 9 times running★ is $(\frac{1}{2})^9 = 1/512$. Although not impossible this is an unlikely result and indicates a distinct, if unlikely, possibility that the coin is two-headed – in which case a 'tail' is very unlikely on the next toss. So the third argument is, on balance, the one most likely to lead to a correct decision – i.e. that on the next toss a 'tail' is less likely to fall than a 'head'.

Note: ★That we are not concerned at this point with the probability that we obtain 10 'heads' in succession but the probability of obtaining the same result 10 times in succession. This means that we are indifferent to the result of the first toss and only interested in the probability of the following 9 being the same.

Progress test 20

1. Here $n = 10$, $p = 0.1$ and $q = 1 - 0.1 = 0.9$

(*a*) (*i*) The binomial distribution formula in this case is:

$$P(x) = \binom{n}{x} p^x q^{n-x} = \frac{10!}{(10-x)!x!} \times 0.1^x \times 0.9^{10-x}$$

The layout, then, is as follows:

x (Houses without bathrooms)	P(x)
0	$\dfrac{10!}{(10-0)! \times 0!} \times 0.1^0 \times 0.9^{10-0} = 0.3487$
1	$\dfrac{10!}{(10-1)! \times 1!} \times 0.1^1 \times 0.9^{10-1} = 0.3874$
2	$\dfrac{10!}{(10-2)! \times 2!} \times 0.1^2 \times 0.9^{10-2} = 0.1937$
3	$\dfrac{10!}{(10-3)! \times 3!} \times 0.1^3 \times 0.9^{10-3} = 0.0574$
4	$\dfrac{10!}{(10-4)! \times 4!} \times 0.1^4 \times 0.9^{10-4} = 0.0112$
5	$\dfrac{10!}{(10-5)! \times 5!} \times 0.1^5 \times 0.9^{10-5} = 0.0015$
6 or more	Balance = $\underline{0.0001}$
	$\underline{\underline{1.0000}}$

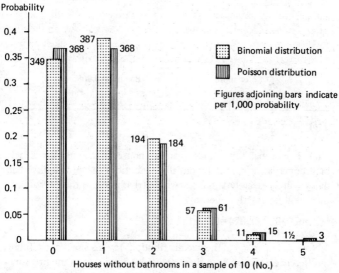

Figure A4.11 *Binomial and Poisson distribution compared when* n = *10 and* p = *0.1*

(*ii*) Here $a = np = 0.1 \times 10 = 1$

The Poisson distribution formula, therefore, is:

$$P(x) = e^{-a} \times \frac{a^x}{x!} = e^{-1} \times \frac{1^x}{x!}$$

Now since $e^{-1} = 0.3679$ (Appendix 3) and 1^x always equals 1 no matter what x might be, then $e^{-1} \times 1^x = 0.3679$. So we have $P(x) = 0.3679/x!$ The layout, then, is:

x (*Houses without bathrooms*)	$P(x)$
0	$\dfrac{0.3679}{0!} = 0.3679$
1	$\dfrac{0.3679}{1!} = 0.3679$
2	$\dfrac{0.3679}{2!} = 0.1840$
3	$\dfrac{0.3679}{3!} = 0.0613$
4	$\dfrac{0.3679}{4!} = 0.0153$
5	$\dfrac{0.3679}{5!} = 0.0031$
6 or more	Balance $= 0.0005$
	$\overline{1.0000}$

(*b*) *See* Fig. A4.11.

(*c*) The probability that less than two houses in the sample will lack bathrooms is (using the binomial distribution) the probability that there will be exactly 0 houses without bathrooms + the probability there will be exactly 1 house

$$= 0.3487 + 0.3874 = \underline{0.7361}$$

Note: Comparison of the two distributions reveals small but definite differences, though even these are under 7 per cent in the first four classes. However, as was observed when the question was set, the situation is a very borderline one for using a Poisson distribution. In more appropriate circumstances the differences would be very much smaller.

2. $a = 2$. $\therefore e^{-a} = 0.1353$.

Let x = number of demands for a coach (when $x > 2$ no coach is available)

$$P(x) = e^{-a} \times \frac{a^x}{x!}$$

x	a^x	$x!$	Probability
0	$2^0 = 1$	1	$0.1353 \times 1/1 = 0.1353$
1	$2^1 = 2$	1	$0.1353 \times 2/1 = 0.2706$
2	$2^2 = 4$	2	$0.1353 \times 4/2 = 0.2706$
>2			Balance $= \underline{0.3235}$
			1.0000

(a) Proportion of days neither coach used $= \underline{0.1353}$

(b) Proportion of days when at least 1 demand refused (i.e. $x > 2$) $= \underline{0.3235}$

(c) Both coaches are out of use for a proportion of 0.1353 of a day, 1 coach out of use for 0.2706 of a day. So one particular coach will be out of use for $0.2706/2 = 0.1353$ of a day.

$\therefore$ Total proportion of a day that one particular coach is out of use = proportion both coaches are out of use + proportion that coach is out of use when only 1 coach is demanded $= 0.1353 + 0.1353 = \underline{0.2706}$

3. $p = 0.8$ $q = 0.2$ $n = 10$.

$$P(x) = \binom{10}{x} \times 0.8^x \times 0.2^{10-x}$$

x	Probability	
7	$10!/((10-7)! \times 7!) \times 0.8^7 \times 0.2^{10-7}$	$= 0.2013$
8	$10!/((10-8)! \times 8!) \times 0.8^8 \times 0.2^{10-8}$	$= 0.3020$
9	$10!/((10-9)! \times 9!) \times 0.8^9 \times 0.2^{10-9}$	$= 0.2684$
10	$10!/((10-10)! \times 10!) \times 0.8^{10} \times 0.2^{10-10}$	$= 0.1074$

(a) Probability that aircraft takes off full is the probability that 8 or 9 or 10 passengers arrive. So:

$$P(x = 8 \cup x = 9 \cup x = 10) = 0.3020 + 0.2684 + 0.1074 = \underline{0.6778}$$

(b) Probability that there are at least 2 empty seats is probability that 6 or less passengers arrive – which is 1 minus probability of 7 or more. So:

$$P(x<7) = 1 - P(x=7 \cup x>7) = 1 - (0.2013 + 0.6778^*) = \underline{0.1209}$$

*As found in (a)

Progress test 21

1. (a) (i) These points are at mean and upper 1σ
∴ Area between = $33\frac{1}{3}$ per cent
∴ Number of items with weights between 115 and 118kg

$$= 33\frac{1}{3} \text{ per cent of } 10,000 = \underline{3,333 \ (3,413^*)}$$

(ii) These points are at lower 1σ and mean.
Number of items = $33\frac{1}{3}$ per cent of 10,000

$$\underline{= 3,333 \ (3,413^*)}$$

(iii) These points are at upper and lower 2σ points, i.e. $115 \pm 2 \times 3$.
∴ Number of items = 95 per cent of 10,000

$$\underline{= 9,500 \ (9,544^*)}$$

(iv) These points are at upper and lower 3σ, i.e. $115 \pm 3 \times 3$.
∴ Number of items = $99\frac{3}{4}$ per cent of 10,000

$$\underline{= 9,975 \ (9,974^*)}$$

(b) 109 to 121 kg is $115 \pm 2 \times 3$, i.e. these weights lie at the 2σ points. Now, 95 per cent of all the distribution falls between these points. Therefore 19 out of 20 items will be between 109 and 121 kg, so any prediction that the weight of an item selected at random will lie between 109 and 121kg can be made with a 95 per cent degree of confidence.

Note: *This is the figure obtained if the table in Appendix 3:**1** is used.

2.

	(i) Undersized rods	(ii) Oversized rods
Acceptance limit	19.90cm	20.10cm
Production mean	20.02cm	20.02cm
Deviation from mean	0.12cm	0.08cm
Standard deviation	0.05cm	0.05cm
Deviations measured in standard deviations (z)	0.12/0.05 = 2.4	0.08/0.05 = 1.6
% rods falling between mean and z (Appendix 3)	49.18	44.52
% items in entire half of distribution	50.00	50.00
% items lying beyond acceptance level	0.82	5.48

So 0.82% of the rods will be rejected as undersized and 5.48% will be rejected as oversized.

3. A rope with a breaking strength of less than 750 kg is no good. Now:

	Supplier A	Supplier B
Mean breaking strength	1,000kg	900kg
Safety margin above 750kg	250kg	150kg
Standard deviation	100kg	50kg
Safety margin in standard deviations	250/100 = 2.5	150/50 = 3

This analysis shows that a bigger proportion of A's ropes fall below 750kg than B's ropes, despite A's ropes having a higher mean. So the company should buy B's ropes, all other things being equal. (Since the area under half the Normal curve is 0.5 then A's ropes fail $0.5000 - 0.4938^* = 0.0062$ of the time and B's ropes $0.5000 - 0.4987 = 0.0013$ of the time.)

*Note: *See* Appendix 3.

4. Mean = 41.2 oz.

(a) $z_1 = \dfrac{40 - 41.2}{0.8} = -1.5.$ So area 40 to mean = 0.4332

$z_2 = \dfrac{42 - 41.2}{0.8} = +1.0.$ So area mean to 42 = 0.3413

∴ proportion boxes between 40 and 42 oz. = 0.7745

(b) 20% = 0.2. Now, if 0.2 of the area of a Normal curve is to fall below a required value then it means that the area between that value and the mean, i.e. the half-curve area *above* the value, must be $0.5 - 0.2 = 0.3$ of the Normal curve area. And the nearest value to 0.3 in Appendix 3:**1** is when $z = 0.85$ (interpolating).

So 20% of the boxes fall below $41.2 - 0.85 \times 0.8 = \underline{\underline{40.52 \text{ oz}}}$

(c) $z = \dfrac{40 - 41.2}{0.8} = -1.5$

So area under the Normal curve which lies below 40 oz. $= 0.5000 - 0.4332 = 0.0668$

∴ out of 100 boxes $100 \times 0.0668 = 6.68$ on average will be scrapped for a cost of £1 $\times 6.68 = \underline{\underline{£6.68}}$

(d) An area of 1% of the Normal curve is a proportion of 0.01.

∴ 0.01 boxes below 40 oz means 0.4900 boxes between 40 oz and the new mean.

For an area to be 0.4900 in the Normal curve tables, z must be about -2.3.

Now $z = \dfrac{x - \bar{x}}{\sigma}$

∴ $-2.3 = \dfrac{40 - \bar{x}}{0.8}$

∴ $\bar{x} = 40 + 0.8 \times 2.3 = 41.84$

So the mean weight of the boxes should be adjusted from 41.2 oz to $\underline{\underline{41.84 \text{ oz}}}$

5. First we use the binomial distribution to find the mean and standard deviation of the distribution of travellers arriving.

Since the airline books 225 people then $n = 225$, and since there is an independent probability of a passenger arriving of 0.8 then $p = 0.8$ (and $q = 0.2$, of course).

So mean of distribution $= np = 225 \times 0.8 - 180$

And standard deviation $= \sqrt{npq} = \sqrt{225 \times 0.8 \times 0.2} = 6$

Since there are a maximum of 195 seats then 15 passengers above the mean can arrive and still all have seats. If 16 or more passengers above the mean arrive, however, then more passengers will have arrived than there are seats.

We next turn to the Normal curve and first it must be pointed out that the Normal curve is a continuous curve whereas our passengers are all discrete. This means that the rule about class limits for grouped frequency distributions has to be applied, i.e. that the limits are extended by ½ unit to create mathematical limits (*see* 7:**9**). So our class of 16 passengers must be taken as a class 15½ to 16½. So, for the purpose of using the Normal curve, we will say that if 15½ more passengers than the mean arrive then more passengers have arrived than seats available. Now, 15½ passengers above the mean is 15½/6 standard deviations, i.e. 2.58 standard deviations. This, then, is the value of z and our Normal curve tables show that this z value relates to an area of about 0.495.

So the probability of more passengers arriving than seats available is $0.500 - 0.495 = \underline{0.005}$

Progress test 22

1. (*i*) $\sigma_{\bar{x}} = \dfrac{15}{\sqrt{25}} = 3\text{kg}$

(*a*) Mean estimate at 95 per cent level

$= 950 \pm 2 \times 3 = \underline{944 \text{ to } 956\text{kg}}$

(*b*) Mean estimate at 99.75 per cent level

$= 950 \pm 3 \times 3 = \underline{941 \text{ to } 959\text{kg}}$

(*ii*) $\sigma_{\bar{x}} = \dfrac{0.08}{\sqrt{100}} = 0.08\text{cm}$

(*a*) Mean estimate at 95 per cent level

$= 1.82 \pm 2 \times 0.08 = \underline{1.66\text{cm to } 1.98\text{cm}}$

(*b*) Mean estimate at 99.75 per cent level

$= 1.82 \pm 3 \times 0.08 = \underline{1.58\text{cm to } 2.06\text{cm}}$

(iii) $\sigma_{\bar{x}} = \dfrac{0.8}{\sqrt{10,000}} = 0.008$cm

   (a) Mean estimate at 95 per cent level

      $= 1.82 \pm 2 \times 0.008 = \underline{1.804\text{cm to } 1.836\text{cm}}$

   (b) Mean estimate at 99.75 per cent level

      $= 1.82 \pm 3 \times 0.008 = \underline{1.796\text{cm to } 1.844\text{cm}}$

Note: the narrower limits that follow the use of a larger sample size.

2. *(a)* $\sigma_p = \sqrt{\left(\dfrac{0.61 \times 0.39}{100}\right)} = 0.0488$

Proportion estimate 95 per cent level
   $= 0.61 \pm 2 \times 0.0488 = 0.512$ to 0.708
   $\simeq \underline{51.2 \text{ per cent to } 70.8 \text{ per cent males}}$

   (b) $\sigma_p = \sqrt{\left(\dfrac{0.61 \times 0.39}{10,000}\right)} = 0.00488$

Proportion estimate 95 per cent level
   $= 0.61 \pm 2 \times 0.00488$
   $\simeq \underline{60.0 \text{ per cent to } 62.0 \text{ per cent males}}$

   (c) $\sigma_p = \sqrt{\left(\dfrac{\dfrac{26}{49} \times \dfrac{23}{49}}{49}\right)} = 0.071$

Proportion estimate 95 per cent level
   $= 26/49 \pm 2 \times 0.071$
   $\simeq \underline{38.8 \text{ per cent to } 67.2 \text{ per cent defectives}}$

Note: When estimating proportions a very much larger sample size is required than is needed when estimating means in order to give estimate limits which are close enough to be worth while.

3. Combined mean $= 28 + 8 = 36$g.
   Combined standard deviation $= \sqrt{(2^2 + 0.75^2)} = 2.14$g.

$\therefore$ Estimate of weight combined nut and bolt is 36g with a standard deviation of 2.14g.

4. If the estimate reads ± 2cm at the 95 per cent level of confidence, then $2\text{cm} = 2 \times \sigma_{\bar{x}}$.

$\therefore \sigma_{\bar{x}} = 1\text{cm}$

Since $\sigma_{\bar{x}} = \dfrac{s}{\sqrt{n}}$, then $1 = \dfrac{s}{\sqrt{100}} = \dfrac{s}{10}$

$\therefore s = 10\text{cm}$

The investigator wants an estimate within $\frac{1}{2}$cm at the 95 per cent level of confidence.

$\therefore \frac{1}{2}\text{cm} = 2 \times \sigma_{\bar{x}}$ (i.e. the improved $\sigma_{\bar{x}}$)

$\therefore \sigma_{\bar{x}} = \frac{1}{4}\text{cm}$

Again, since $\sigma_{\bar{x}} = \dfrac{s}{\sqrt{n}}$, $\dfrac{1}{4} = \dfrac{10}{\sqrt{n}}$

$\therefore \sqrt{n} = 40$

$\therefore n = 1{,}600$

$\therefore$ The sample size will need to be $\underline{1{,}600}$

5. If a 99 per cent confidence level is required, it entails finding how far either side the mean one has to measure in order to include 99 per cent of the total area beneath the Normal curve or, in other words how far *one side* of the mean one has to measure to include $99/2 = 49.5$ per cent of the total area. Looking up the table in Appendix 3, it can be seen that an area of 0.4953 is embraced by using 2.6σ. Therefore $\sigma_{\bar{x}}$ needs to be multiplied by 2.6 to give a confidence level of 99 per cent.

6. Null hypothesis: there is no contradiction between a population mean of 14 hours and a sample mean of 13 hours 20 minutes.

Test: Two-tail.

Now $\sigma_{\bar{x}} = \dfrac{s}{\sqrt{n}} = \dfrac{3}{\sqrt{64}} = \underline{\underline{\frac{3}{8} \text{ hour.}}}$

If 14 hours is the population mean, then 95 per cent sample means will fall between $14 \pm 2 \times \frac{3}{8} = 13\frac{1}{4}$ and $14\frac{3}{4}$ hours.

The sample mean is $13\frac{1}{3}$ hours and therefore lies between these limits.

Therefore the null hypothesis cannot be rejected at the 5 per cent level of significance, and the welfare officer's assertion is not disproved.

7. Null hypothesis: there is no contradiction between a population proportion of 0.6 and a sample proportion of $1{,}410/2{,}500 = 0.564$.

Test: Two-tail (the candidate has *not* asserted that *at least* 60 per cent of the voters support him).

Now if $p = 0.6^*$, $q = 1 - 0.6 = 0.4$, and $n = 2,500$, then:

$$\sigma_p = \sqrt{\left(\frac{pq}{n}\right)} = \sqrt{\left(\frac{0.6 \times 0.4}{2,500}\right)} = 0.0098$$

If 0.6 is the population proportion, then 99 per cent sample proportions will fall between $0.6 \pm 2.6 \times 0.0098 = 0.5745$ and 0.6255.

The sample proportion is 0.564, which is *outside* these limits, hence the null hypothesis must be rejected at the 1 per cent significance level, i.e. there *is* a contradiction between the asserted population proportion and the sample proportion.

This means that the population proportion is not 60 per cent.**

Notes:

* Remember: use the true population proportion for p where possible (*see* **36**). Although the true proportion is not known here, it represents the true proportion until such time as it is proved untrue.

† Note how the size of the sample improves our discrimination. The candidate claimed in effect that 1,500 out of every 2,500 people supported him. The sample gave 1,410: a mere 90 short, on a sample of perhaps only 3 per cent or 4 per cent of the total voters. Nevertheless, with a sample of 2,500 we were able to say that his assertion was wrong, and say this, moreover, knowing we would only make an error of wrongly dismissing such a claim once in 100 times. (We are, of course, assuming the people sampled told the truth!)

8. (a) 30 per cent and 40 per cent are equivalent to proportions of 0.3 and 0.4 respectively.

$$\sigma p = \sqrt{pq/n} = \sqrt{0.3 \times 0.7/100} = 0.046$$

Now, for 40 per cent or more defectives $z = \dfrac{0.4 - 0.3}{0.046} = 2.174$.

From the Normal curve tables this z indicates that the area between 0.3 and 0.4 is approximately 0.485.

∴ probability of 40 per cent or more defectives $= 0.500 - 0.485 = \underline{\underline{0.015}}$

(b) Since we are not interested in whether or not the process produces less than 30 per cent defectives the test we need is a one-tail test.

Now, for 42 or more defectives in 100 items $z = \dfrac{0.42 - 0.3}{0.046} = 2.61$.

This z indicates that the area between 0.30 and 0.42 is approximately 0.4955.

∴ probability of 42 or more defectives in 100 items $= 0.500 - 0.4955 = 0.0045$ i.e. less than 1 in 200. This strongly suggests that the claim is not valid. However, it is better to establish a level of significance *before* making a significance test and so this should be done next, after which a further sample should be tested.

9. If the two samples were taken from the same population, it would be equivalent to two samples taken from two populations having the same mean. Therefore, to test the expert's assertion, we need to find out whether the difference between the means is significant.

Now $\sigma_{\bar{x}_1} = \dfrac{2}{\sqrt{200}}$ and $\sigma_{\bar{x}_2} = \dfrac{2}{\sqrt{80}}$.

Since $\sigma_{(\bar{x}_1 - \bar{x}_2)} = \sqrt{(\sigma^2_{\bar{x}_1} + \sigma^2_{\bar{x}_2})}$ then

$$\sigma_{(\bar{x}_1 - \bar{x}_2)} = \sqrt{\left(\left(\frac{2}{\sqrt{200}}\right)^2 + \left(\frac{2}{\sqrt{80}}\right)^2 \right)} = \sqrt{\left(\frac{4}{200} + \frac{4}{80}\right)}$$

$$= \underline{0.264\text{kg}}$$

Therefore 95 per cent of the differences between means can extend to $2 \times \sigma_{(\bar{x}_1 - \bar{x}_2)} = 2 \times 0.264 = 0.528$kg.

The actual difference is $20\frac{1}{2} - 20 = 0.5$kg. This is within the allowed limits and there is, therefore, no evidence at the 5 per cent level of significance to disprove the expert's assertion that the two samples were taken from a single population.

10. This question calls for a one-tail test of significance of the difference between proportions (one-tail, since the question is not as to whether there is a difference between the fitness of the citizens of the two cities but as to whether there is a difference showing that citizens of A specifically are fitter than those of B).

Here the proportion of citizens passing in A is $96/200 = 0.48$ and the proportion of citizens passing in B is $84/200 = 0.42$.

$$\therefore \quad \sigma_{p_A} = \sqrt{\left(\frac{pq}{n}\right)} = \sqrt{\left(\frac{0.48 \times 0.52^\star}{200}\right)}$$

and $$\sigma_{p_B} = \sqrt{\left(\frac{0.42 \times 0.58^\star}{200}\right)}$$

Since $\sigma_{(p_1-p_2)}=\sqrt{(\sigma^2_{p_A}+\sigma^2_{p_B})}$

then $\sigma_{(p_1-p_2)}=\sqrt{\left(\left(\sqrt{\left(\dfrac{0.48\times0.52}{200}\right)}\right)^2+\left(\sqrt{\left(\dfrac{0.42\times0.58}{200}\right)}\right)^2\right)}$

$=\sqrt{\left(\dfrac{0.48\times0.52}{200}+\dfrac{0.42\times0.58}{200}\right)}=0.04966$

Now a 5 per cent one-tail level of significance is the same as a 10 per cent two-tail level of significance. The table in Appendix 3 shows that a z of 1.65 (interpolating) will give an area 0.45 of the curve, i.e. 90 per cent when doubled. So 95 per cent of the differences between proportions, where one given proportion is greater than the other, do not exceed 1.65 standard errors. Our 95 per cent confidence limit, then, is $1.65\times0.04966=0.0819$.

And the actual difference between the proportions is $0.48-0.42 =0.06$. Since this difference is within the allowed limits it is not significant.

This means that the difference between the 96 fit citizens of A and the 84 fit citizens of B could have arisen by chance alone and the health official's claim that the citizens of A are fitter than those of B is not proven.†

Notes:

* It pays not to simplify these expressions, since at the next stage they are squared, and that means that one has only to remove the square root sign.

†Note how a small sample in a proportion problem can mean that quite large differences can occur without the differences being significant (thus, in this case, a difference of $96-84=12$ in a sample size of 200, i.e. 6 per cent, is not significant).

11. Assume that it is asserted that 60 per cent of a population are males. You have taken a random sample of 10 people and found that 3 only are males. If you are only interested in a result where the percentage of males in the population falls to 30 or under in a sample of 10, you can test if this result is statistically significant at the 5 per cent level of significance by the following application of the binomial distribution.

Binomial distribution of number of males in a sample of 10 if proportion of males in the population is 0.6:

No. Males	Probability	Combined probabilities
0	$\binom{10}{0} \times 0.6^0 \times 0.4^{10} = 0.0001$	
1	$\binom{10}{1} \times 0.6^1 \times 0.4^9 = 0.0016$	
2	$\binom{10}{2} \times 0.6^2 \times 0.4^8 = 0.0106$	0.0548
3	$\binom{10}{3} \times 0.6^3 \times 0.4^7 = 0.0425$	
4	$\binom{10}{4} \times 0.6^4 \times 0.4^6 = 0.1115$	
5	$\binom{10}{5} \times 0.6^5 \times 0.4^5 = 0.2007$	
6	$\binom{10}{6} \times 0.6^6 \times 0.4^4 = 0.2508$	0.6665
7	$\binom{10}{7} \times 0.6^7 \times 0.4^3 = 0.2150$	
8	$\binom{10}{8} \times 0.6^8 \times 0.4^2 = 0.1209$	
9	$\binom{10}{9} \times 0.6^9 \times 0.4^1 = 0.0403$	
10	$\binom{10}{10} \times 0.6^{10} \times 0.4^0 = 0.0060$	
	1.0000	

As can be seen, the probability of selecting a sample which contains 3 or less men when the population contains 60 per cent males is 5.48 per cent – which is just above one in twenty. So the null hypothesis cannot quite be rejected in this instance.

This example illustrates the relevance of the binomial distribution to sampling theory. In practice, however, much larger samples than 10 are usually taken and then the arithmetic grows in complexity. Fortunately with larger samples the normal distribution approximates to the binomial distribution and can be used in lieu.

Note: The full distribution has been given here so that the student can see that even with a sample as small as 10, when $p=0.6$ the distribution is not far from being symmetrical around the peak of 6. Moreover, the mean and standard deviation of the binomial distribution are np and $\sqrt{(npq)}$, i.e. 6 and $\sqrt{(10 \times 0.6 \times 0.4)} = 1.55$. In a Normal distribution 68.26 per cent of the distribution lies within the 1σ limits. Here these limits would be $6 \pm 1.55 \simeq 4\frac{1}{2}$ and $7\frac{1}{2}$ – which means that the discrete numbers 5, 6 and 7 would collectively have a 0.6826 probability of occurring. As can be seen from above using the binomial distribution leads to a probability of 0.6665. So even at this small size of sample the normal distribution is already approximating quite closely to the binomial distribution (though if p were nearer 1 or 0 the approximation would not be quite so close).

Appendix 5
Examination technique

To pass any examination you must:

(a) have the knowledge;
(b) convince the examiner you have the knowledge;
(c) convince him within the time allowed.

In the book so far we have considered the first of these only. Success in the other two respects will be much more assured if you apply the examination hints given below.

1. Answer the question. Apart from ignorance, *failure to answer the question is undoubtedly the greatest bar to success.* No matter how often students are told, they always seem to be guilty of this fault. If you are asked for a frequency polygon, *don't* give a frequency curve; if asked to give the features of the mean *don't* detail the steps for computing it. You can write a hundred pages of brilliant exposition, but if it's not in answer to the set question you will be given no more marks than if it had been a paragraph of utter drivel. To ensure you answer the question:

(a) *read the question carefully;*
(b) *decide what the examiner wants;*
(c) *underline the nub of the question;*
(d) *do just what the examiner asks;*
(e) *keep returning to the question as you work on the answer to ensure that you are still answering it.*

2. Put your ideas in logical order. It is quicker, more accurate and gives a greater impression of competence if you follow a predetermined logical path instead of jumping about from place to place as ideas come to you.

3. Maximise the points you make. Examiners are more impressed by a solid mass of points than an unending development of one solitary idea, no matter how sophisticated and exhaustive. Do not allow yourself to become bogged down with your favourite hobby-horse.

4. Allocate your time. The marks allocated to questions often bear a close relationship to the time needed for an appropriate answer. Consequently the time spent on a question should be in proportion to the marks. Divide the total exam marks into the total exam time (less planning time) to obtain a 'minutes per mark' figure, and allow that many minutes per mark of each individual question.

5. Attempt all questions asked for. Always remember that the first 50 per cent of the marks for any question are the easier to earn. Unless you are working in complete ignorance, you will always earn more marks per minute while answering a new question than while continuing to answer one that is more than half done. So you can earn many more marks by half-completing two answers than by completing either one individually.

6. Do not show your ignorance. Concentrate on displaying your knowledge, not your ignorance. There is almost always one question you need to attempt and are not happy about. In answer to such a question put down all you *do* know, and then devote the unused time to improving some other answer. Certainly you will not get full marks by doing this, but neither will you fill your page with nonsense. By spending the saved time on another answer you will at least be gaining the odd mark or so.

7. If time runs out. What should you do if you find time is running out? The following are the recommended tactics.

(*a*) If it is a mathematical answer, do not bother to work out the figures. Show the examiner by means of your layout that you know what steps need to be taken and which pieces of data are applicable.

He is very much more concerned with this than with your ability to calculate.

(*b*) If it is an essay answer, put down your answer in the form of notes. It is surprising what a large percentage of the question marks can be obtained by a dozen terse, relevant notes.

(*c*) Make sure that every question and question part has some answer – no matter how short – that summarises the key elements.

(*d*) Do not worry. Shortage of time is more often a sign of knowing too much than too little.

8. Avoid panic, but welcome 'nerves'. Nerves are a great aid in examinations. Being nervous enables one to work at a much more concentrated pitch for a longer time without fatigue. Panic, on the other hand, destroys one's judgement. To avoid panic:

(*a*) know your subject (this is your best 'panic-killer');

(*b*) give yourself a generous time allowance to read the paper – quick starters are usually poor performers;

(*c*) take two or three deep breaths;

(*d*) concentrate simply on maximising your marks – leave considerations of passing or failing until after.

(*e*) answer the easiest questions first – it helps to build confidence;

(*f*) do not let first impressions of the paper upset you. Given a few minutes, it is amazing what one's subconscious will throw up. Moreover it is often only the unfamiliar presentation of data that makes a statistical question look difficult: once you have thought it through carefully, it often shows itself to be quite simple.

Appendix 6
Examination questions

1. Badly worded statements can bring the subject of statistics into disrepute.

You are required to consider the following statements and:

(a) explain briefly where they mislead or fail to make sense;

(b) reword them in a more acceptable form.

(i) 'Nine out of ten people in this country would oppose a policy of state intervention in the z industry.'

(ii) 'Unemployment up 10 per cent' . . . as stated in newspaper A, 'Unemployment down 10 per cent' . . . as stated in newspaper B, both on the same day.

(iii) 'There are 2.41 children per family in the county of Y.'

(iv) '80 per cent of car accidents occurred within three miles of the driver's home, therefore, longer journeys must be safer.'

(CIMA)

2. You are required to state and explain briefly any statistical/mathematical weaknesses in five of the following:

(a) In the ABC company the correlation coefficient between production costs and quantity of toys produced in the first forty weeks of 1984 was $r = 0.61$. This shows that 61 per cent of the variation in production costs can be explained by the production level.

(b) A financial journalist has estimated that the odds are 2:1 (i.e. the probability is 2/3) that Ruritania's economy will improve next year, 3:1 that it will *not* remain the same and 5:1 that it will improve or remain the same.

(c) In a certain country the retail price index (January 1980 = 100) changed from 160 in April 1984 to 180 in October 1984. A newspaper reported this fact with the headline 'Retail prices show an average increase of 20 per cent in six months'.

(d) In Ruritania the crude death rate (total deaths per 1,000 of the population) in Town A is 0.012 which is the same as the national figure. The crude death rate in Town B is 0.024. Town B is a dangerous place.

(e) There is at present approximately 80 per cent excess capacity on existing air passenger services between Town C and Town D. In other words, throughout the year the total capacity provided by current air passenger services is used to the extent of only 20 per cent.

(f) As every gambler knows, it is fatal to stop when you are losing since it deprives you of the statistical probability that your losses will be offset, at least partly, by subsequent winnings.

(CIMA)

3. (a) Define:

   (i) absolute error;

   (ii) relative error;

   (iii) biased and unbiased errors.

(b) A business estimated that:

   (i) sales of the year in its four sales areas would be:

Area No:	£
1	850,000
2	1,100,000
3	1,300,000
4	950,000

Calculate, if the estimates were correct to ±£10,000 in each area, the maximum absolute and relative errors possible when considering the total value of sales of the business for the year.

   (ii) Cost of producing the sales with the corresponding estimated errors would be as follows:

	£	%
Wages	1,300,000	±4
Materials	1,600,000	±5
Fixed expenses	800,000	±3

Estimate the profit, stating the limits of possible errors.

(CIMA)

4. The following table is given:

Sri Lanka
Gross domestic product by origin (millions of rupees at current prices)

	1972	1973	1974
Agriculture	4,119	5,025	8,356
Wholesale and retail trade	1,986	2,455	2,560
Manufacturing	1,728	2,017	2,475
Transport, storage, communications	1,333	1,525	1,683
Construction	711	802	1,011
Public administration and defence	575	654	704
All other	2,355	2,787	3,017
	12,807	15,265	19,806

Source: Central Bank of Ceylon, Extracted from Europa year book, Vol. II

You are required to:

(*a*) Approximate the above data to the nearest 1,000 million rupees.

(*b*) Estimate the absolute and percentage relative errors of the total obtained in (*a*) above.

(*c*) Explain why the percentage relative error is a superior measure to the absolute error.

(*d*) Define a systematic error. Explain your answer by redrafting one of the columns in your answer to (*a*) above.

(*e*) Comment on:

(*i*) the likelihood that the increase shown from 1972 to 1974 is a real volume increase;

(*ii*) the classification of the items in the table.

(*f*) Illustrate and explain what is meant by the term significant digits.

(*CIMA*)

5. (*a*) The following table shows the breakdown of a population into categories of sex, age and whether a smoker or non-smoker (frequencies are in thousands).

	Age		
	21–40	*41 and over*	*Total*
Male:			
Smoker	61	45	106
Non-smoker	41	27	68
Female:			
Smoker	68	56	124
Non-smoker	59	43	102

(*i*) It is proposed to construct a quota sample of 100 individuals which fully reflects the distribution of these three characteristics in the population. Calculate the numbers in each category of the quota sample.

(*ii*) Suppose that it is decided after all that age is no longer of importance in the survey. How would your quota sample of 100 be affected?

(*iii*) What is the major criticism of a quota sample?

(*b*) Eighty companies are listed in order of profitability and you wish to take a random sample of twenty of these companies. You are provided with a list of randomly generated digits, a section of which is given below:

46819037154583502

Explain how you could use these random digits to select twenty companies from the complete list and show how the first three companies would be selected using your procedure.

(*ACA*)

6. A survey was made of the proportion of time between 8.00 a.m. and 1.00 p.m. that the average housewife in an area spends working on household duties. (A housewife is defined as a wife who does not have a regular job.) The survey was done by activity sampling, whereby interviewers called at houses and asked the housewife whether she was actually working or not at the moment when the interviewer arrived. Responses to 419 calls were as follows:

| | Hour during which visit took place | | | | | |
	8–9	9–10	10–11	11–12	12–1	Total
Housewife working	21	21	54	36	9	141
Housewife not working	12	15	39	27	9	102
No information (no reply or housewife not at home)	18	24	60	42	32	176
						419

It was concluded from the survey that between 8.00 a.m. and 1.00 p.m., housewives are busy in the house for around 58 per cent (141/243) of the time.

(*a*) Give the main reasons why the conclusions are likely to be biased.

(*b*) During which hour do the figures suggest that housewives are busiest?

(*c*) Ignoring the effect of bias in the sample (that is, assuming a random sample) calculate 95 per cent confidence limits for the estimated proportion of 0.58. Explain briefly the meaning of your calculated values.   (*ACA*)

7. 'In the period 19–0 to 19–4 there were 1,775 stoppages of work arising from industrial disputes on construction sites. Of these 677 arose from pay disputes, 111 from conflict over demarcation, 144 on working conditions and 843 were unofficial walk-outs. Ten per cent of all stoppages lasted only one day; the corresponding proportions for pay disputes was 11 per cent, 6 per cent in the case of demarcation and 12.5 per cent in the case of working conditions.

In the subsequent period 19–5 to 19–9 the number of stoppages was 2,807 higher. But, whereas the number of stoppages arising from pay disputes fell by 351 and those concerning working conditions by 35, demarcation disputes rose by 748. The number of pay disputes resolved within one day decreased by 45 compared with the earlier period, but only by 2 in the case of working conditions. During the later period 52 stoppages over demarcation were settled within the day, but the number of stoppages from all causes which lasted only one day was 64 greater in the later period than in the earlier.'

You are required to:

(a) tabulate the information in the above passage, making whatever calculations are needed to complete the entries in the table, and

(b) write a brief comment on the main features disclosed by the completed table. (CA)

8. The table below refers to some of the results of an investigation involving 76 companies that took part in a CIMA-sponsored research study. This study identified the areas of application of forecasting techniques and how important they were thought to be (i.e. priority) within the companies.

Area of application	Priority	Weekly	Monthly	Quarterly	Annual	3+years	Total
Application of forecasting techniques *(Percentage analysis by timescale)*							
Market size	4	–	6	17	55	22	100
Market share	8	–	14	27	45	14	100
Production planning	5	24	31	17	22	6	100
Stock control	6	11	34	25	21	9	100
Cash planning	1	7	34	22	26	11	100
Research/ development	9	–	7	18	57	18	100
Investment appraisal	7	–	15	21	49	15	100
Profit planning	2	–	21	26	40	13	100
Capital employed	3	–	18	21	46	15	100
Total %		5	22	22	38	13	100

Source: McHugh and Sparkes, Management Accounting, March 1983

State three main conclusions which can be drawn from these data, and illustrate each of these with a suitable summary table or diagram. (CIMA)

9. A printer is investigating how the cost of setting up the printing of books varies with a number of factors. He believes that the principal factors are the size of the book and its degree of technical complexity. The first of these two is measured by the number of equivalent standard pages in the book. For the second the book is classified either as novel (N), technical (T) or mathematical (M), which are in increasing order of printing complexity.

Details were obtained of the set-up cost of producing 22 books of varying size and complexity and these are given below:

Complexity	Number of standard pages	Set up cost (£)	Complexity	Number of standard pages	Set up cost (£)
T	362	241	M	196	143
T	264	174	T	363	234
M	420	384	T	418	246
N	437	207	N	255	133
M	374	334	N	582	247
N	242	121	M	512	412
N	339	161	T	300	161
T	285	195	M	273	256
N	362	182	M	269	277
M	183	176	N	482	199
T	451	260	T	492	296

(*a*) Represent the above data on a graph designed to show how the cost varies with complexity and book length, and comment on any observed features of the graph.

(*b*) Estimate from your graph the approximate increase in set-up cost per 100 pages for mathematical books.

(*c*) Estimate from your graph the set-up cost of a proposed book of length 550 pages whose type complexity is midway between technical and mathematical. (*ACA*)

10. The Central Statistical Office, Financial Statistics in August 1979 contained the following:

Average of daily telegraphic transfer per £1 sterling in London

Date	Swiss francs	Deutsche mark
1977:		
January	4.2701	4.102
July	4.1543	3.934
1978:		
January	3.8398	4.094
July	3.4143	3.892
1979:		
January	3.3479	3.708
July	3.7205	4.122

(a) Using the standard graph paper provided, plot the value of £1 in Swiss and German currencies on a semi-logarithmic scale.

(b) Interpret the graph.

(c) Comment on the merits and demerits of a semi-logarithmic graph compared with a linear scale graph with particular reference to the data given.

Marks will be awarded for good presentation.   (*CIMA*)

11. As management accountant of Z Limited, you are given the following statistics relating to its production of a standard electrical component, made for the domestic market.

Year	Saleable output in thousands	Average number of assembly operators	Average daily output per operator	Hourly earnings per operator in pence
19–1	362	62	26	55
19–2	358	60	26	58
19–3	366	60	27	62
19–4	365	56	28	67
19–5	370	55	29	74
19–6	367	52	31	90

You are required to:

(a) (i) convert all information given to a comparable basis using index numbers;

(ii) construct a ratio-scale (semi-logarithmic) graph and plot your results.

(b) report trends shown in your graph in a brief memorandum to the management;

(c) contrast linear scale graphs with ratio-scale graphs with regard to:

(i) the slope of the curve;
(ii) suitability for analysing an aggregate into its constituents;
(iii) negative values;
(iv) base-line;
(v) suitability of the *y*-axis (vertical) over a great range.

(*CIMA*)

12.

Share of total personal incomes UK 1973–4

Income group	Before income tax %	After income tax %
Top 10%	24.9	21.8
11–20%	14.1	14.3
21–30%	12.1	12.3
31–40%	10.5	10.9
41–50%	9.4	9.7
51–60%	8.1	8.5
61–70%	7.0	7.3
71–80%	5.8	6.2
81–90%	4.6	5.0
91–100%	3.5	4.0

Source: Royal Commission on the Distribution of Income & Wealth

You are required to:

(*a*) illustrate by means of a graphical technique such as the Lorenz curve the degree of inequality in income distribution before and after tax; and

(*b*) discuss the usefulness of the graphical technique you have chosen and state the conclusion you draw from the graph.

(*CA*)

13. Recently the following information was published by a large retail group.

Year	Sales £m	Pre-tax profits £m	Net capital employed £m	Number of employees
19–8 to 19–9	888	57.25	615.6	41,500
19–7 to 19–8	823	52.61	307.5	41,770
19–6 to 19–7	724	46.52	279.7	43,867
19–5 to 19–6	665	40.59	268.8	47,498
19–4 to 19–5	573	36.10	268.6	49,121

(*a*) As management accountant you are to select any set of figures or combination of figures given above and present them in a visual display form. The objective of your display is to aid employees to understand the information presented.

The following information is also given:

General index of retail prices

Year	Index	Increase on previous year %
19–4	100.0	24.2
19–5	116.5	16.5
19–6	135.0	15.8
19–7	146.2	8.3
19–8	165.8	13.4

Marks will be awarded for effective and imaginative use of the information provided.

(*b*) Explain why you consider your presentation should help the reader understand the figures more easily.   (*CIMA*)

14. The management of an industrial company is considering methods of supplying financial information to its employees, many of whom are unfamiliar with financial matters. It is considered that a visual display in chart form, accompanied by oral explanations, is the best method of communication.

The following information for the five years from 19–1 to 19–5 inclusive is available:

	19–5	19–4	19–3	19–2	19–1
Index of price changes	152	145	121	110	100
Financial results in millions of £s					
Sales	6.3	6.5	5.8	4.7	3.9
Direct materials	3.1	3.2	2.8	2.2	1.9
Direct wages	1.4	1.2	1.1	1.0	0.8
Production overhead	1.0	1.0	0.9	0.8	0.6
Other overhead	0.4	0.4	0.3	0.2	0.2
Taxation	0.3	0.3	0.3	0.2	0.2
Profit	0.1	0.4	0.4	0.3	0.2

Product groups: of total
sales

	%	%	%	%	%
Product:					
A	24	22	20	20	15
B	12	10	11	9	12
C	16	21	23	26	30
Other products	48	47	46	45	43
	100	100	100	100	100

As management accountant, prepare for the managing director three
different visual displays as follows:

(*a*) sales by product groups;
(*b*) analysis for 19–5 and 19–4 of total sales into costs, taxation and
profit with particular emphasis on direct wages;
(*c*) total sales year by year with figures adjusted for inflation.

(*CIMA*)

15. The 1977 annual report of Tate and Lyle Limited contained the
following information:

(*a*) 1977 Profits by sector (before central expenses):

Activity	%
Sugar refining	26
Commodity handling, trading, storage and distribution	37
Raw sugar production	4
Starch	8
Engineering, construction materials and miscellaneous group activities	16
Shipping	9

(*b*) Group statement of value added for the year ended 30
September 1977, with comparative figures for year 1976 (all figures in
£ million):

	1977	1976
Value added	175.2	146.1

Applied in the following way:

To pay providers of capital:

Interest on loans	16.5		8.2
Dividends to stockholders	7.3		6.5
Share of minority interests	3.1		2.8
		26.9	17.5

To pay government:

Taxes on profits		13.1	16.9

To pay employees:

Wages, pensions, etc.		97.8	73.6

To provide investment for growth:

Depreciation	17.0		11.8
Retained profits	20.4		20.1
		37.4	31.9
To provide for extraordinary items		–	6.2
		175.2	146.1

You are required to present the information at (*a*) and (*b*) above in different visual display forms to make it more readily understandable by employees. The display for value added (*b*) should show both absolute and proportionate changes when the results for 1976 and 1977 are compared. (*CIMA*)

16. Percentage of owners with assets covered by estate duty statistics.

		Percentages	
Over (£)	Not over (£)	1962	1973
–	1,000	48.3	18.8
1,000	3,000	29.5	25.1
3,000	5,000	10.9	11.9
5,000	10,000	6.2	21.3
10,000	25,000	3.4	17.9
25,000	100,000	1.5	4.4
100,000	–	0.2	0.6
Total number of owners (000s)		18,448	19,140

Source: Social Trends 1976

You are required to:

(a) derive by graphical methods the median, lower and upper quartile values for 1962 and 1973;

(b) calculate for the two distributions the median values and explain any difference between these results and those obtained in (a);

(c) estimate for each year the number of wealth owners with less than (i) £2,500 and (ii) £25,000 and comment on the reliability of your estimates; and

(d) write a brief report on the data, using your results where relevant. (*CA*)

17. The following table shows the income distribution of the middle management of a large industrial company.

Incomes

Range £	Number of managers
2,950 and <3,150	10
3,150 and <3,350	8
3,350 and <3,550	18
3,550 and <3,750	22
3,750 and <3,950	23
3,950 and <4,150	25
4,150 and <4,350	35
4,350 and <4,550	30
4,550 and <4,750	14
4,750 and <4,950	6
4,950 and <5,150	5
5,150 and <5,350	4

You are required to:

(a) prepare a cumulative frequency curve of the above distribution;

(b) (i) calculate the first and third quartile salaries, and the median salary; and

(ii) show them on the graph prepared for part (a);

(c) give a brief note on a business use to which this form of presentation may apply. (*CIMA*)

18. The management of a manufacturing company is considering using aspects of statistical techniques in its budgeting programme.

(a) Four products, Tweed, Tay, Tees and Tyne, are produced. The following information is available to the management accountant:

	Labour standard hours per thousand	Units produced in thousands		
		October	November	December
Tweed	8	51	47	35
Tay	3	16	21	16
Tees	5	18	29	15
Tyne	7	30	31	23
Number of working days		21	20	16

You are required to calculate a productivity index for either November or December with October as the base.

(b) The sales manager considers that the modal value of his salesmen's weekly journeys would provide him with the most useful average for budgeting purposes. The following distribution is given:

Miles travelled	Number of salesmen
100 and <200	3
200 and <300	5
300 and <400	2
400 and <500	8
500 and <600	2

You are required to calculate the modal value and to advise the sales manager of the consequences of using this result.

(c) The sales manager's team averaged the following monthly order values during the last budget period:

Orders obtained
Average for month: last budget period

Value in £000	Number of salesmen
18 and <22	1
22 and <26	3
26 and <30	2
30 and <34	5
34 and <38	4
38 and <42	3
42 and <46	2

You are required to:

(*i*) compute the monthly median value of orders received;

(*ii*) evaluate its use for budgeting purposes.   (*CIMA*)

19. An international company, reviewing the orders received by its European sales outlets, has compiled the following information:

Value of orders received July

Monthly value of orders (£000)	Number of sales outlets
180 and <220	10
220 and <260	30
260 and <300	20
300 and <340	50
340 and <380	40
380 and <420	30
420 and <460	20

The management wishes to establish an average monthly order value which it can use in budgeting and fixing discounts for the outlets.

You are required to:

(*a*) state the type of average you would recommend management to use, giving briefly reasons for your choice;

(*b*) calculate the average order value using the basis you have recommended;

(*c*) comment on the superiority of your recommended basis over alternatives that may have been suggested. (*CIMA*)

20. In a community there are f_0 families with no children, f_1 families with 1 child, f_2 with 2 children and so on, the largest family size being 8 children. A sum of money, M, is to be allocated amongst these families on the basis of a fixed amount per child in the family.

(*a*) Give expressions for

(*i*) the amount allocated per child
(*ii*) the mean amount received per family.

(*b*) Suppose that the quantities $f_0, f_1, f_2 \ldots f_8$ are arranged in ascending order of magnitude as

$$f_7 < f_8 < f_6 < f_5 < f_1 < f_4 < f_0 < f_3 < f_2$$

Give an expression for the modal amount of money received per family.

(*c*) You are now given that $f_0 = 72, f_1 = 30, f_2 = 142, f_3 = 105, f_4 = 41, f_5 = 7, f_6 = 2, f_7 = 0, f_8 = 1, M = £10,000$.

Using your expression derived in parts (*a*) (*i*), (*ii*) and (*b*) above, or by some alternative method, find numerical values for the amount per child, mean amount per family and modal amount per family. Find also the mean number of children per family. (*ACA*)

21. A small supermarket operates one checkout point with a cashier. Customers are checked out one at a time on a first-come, first-served basis. Checkout times vary, depending on the quantity of items bought. As soon as one customer is checked out, the cashier is free to deal with the next.

There are periods when customers queue, waiting for the cashier to become free, and periods when the cashier is free, waiting for a customer to arrive. The data on page 422 relates to the first thirteen customers checked out one morning.

(*a*) Show how to determine from the data the checkout time of a customer, and illustrate the variability in the above checkout times by means of a frequency distribution.

Customer	Arrival time at checkout	Checkout completion time
1	9.05	9.08
2	9.11	9.12
3	9.12	9.16
4	9.15	9.20
5	9.22	9.25
6	9.23	9.28
7	9.27	9.33
8	9.30	9.37
9	9.36	9.39
10	9.39	9.44
11	9.42	9.48
12	9.45	9.52
13	9.53	9.55

(*b*) From your distribution derived in (*a*), calculate:

(*i*) the sample mean checkout time per customer;
(*ii*) the sample standard deviation of checkout time.

(*c*) An early morning customer observes that there are six customers ahead of him in the queue, including the one at the front whose checkout is just about to start. Estimate the mean and standard deviation of the time that he will wait before his own checkout is complete. (*ACA*)

22. Items are produced to a target dimension of 3.25cm on a single machine. Production is carried out on each of three shifts, each shift having a different operator, A, B and C. It is decided to investigate the accuracy of the operators. The results of a sample of 100 items produced by each operator are as follows:

Operator	Dimension of item (cm)								
	3.21	3.22	3.23	3.24	3.25	3.26	3.27	3.28	3.29
A	1	6	10	21	36	17	6	2	1
B	0	0	7	25	34	27	6	1	0
C	3	22	49	22	4	0	0	0	0

(a) State, with brief explanation, which of the operators is

(i) the most accurate in keeping to the target;
(ii) the most consistent in his results.

(b) Illustrate your answers to (a) by comparing the results obtained by each operator as frequency polygons on the same graph.

(c) Why is it that a measure of the mean level of the dimension obtained by each operator is an inadequate guide to his performance?

(d) Calculate the semi-interquartile range of the results obtained by operator A, to 3 decimal places, and comment briefly on the usefulness of this calculation.   (ACA)

23. (a) Using the figures given below calculate:

(i) the range;
(ii) the arithmetic mean;
(iii) the median;
(iv) the lower quartile;
(v) the upper quartile;
(vi) the quartile deviation;
(vii) the mean deviation;
(viii) the standard deviation.

2	15	30	43	55
5	17	32	45	58
7	18	36	47	60
8	22	39	51	64
11	26	40	53	66

(b) Explain the term 'measure of dispersion' and state briefly the advantages and disadvantages of using the following measures of dispersion:

(i) range;
(ii) quartile deviation;
(iii) mean deviation;
(iv) standard deviation.   (CIMA)

24. An air charter company has been requested to quote a realistic turn-round time for a contract to handle certain imports and exports of a fragile nature.

The contracts manager has provided the management accountant with the following analysis of turn-round times for similar goods over a given twelve-month period.

Turn-round time in hours	Number of frequencies
Less than 2	25
2 and < 4	36
4 and < 6	66
6 and < 8	47
8 and <10	26
10 and <12	18
12 and <14	2

You are required to:

(a) calculate from the distribution give above:

(i) the mean;
(ii) the standard deviation;

(b) advise the contracts manager of the turn-round time to be quoted using:

(i) the mean plus one standard deviation;
(ii) the mean plus two standard deviations;

stating in each case the chance of success in meeting the turn-round time given;

(c) explain the merits and the demerits of the standard deviation as a form of measure.   (CIMA)

25. The following distribution was extracted from a report prepared by the quality control department of a manufacturing company:

Rejects (units of output)	Number of batches
13 to 17 inclusive	20
18 to 22	30
23 to 27	30
28 to 32	32
33 to 37	28
38 to 42	22
43 to 47	12
48 to 52	6

You are required to:
(a) calculate from the distribution given above:

(i) the mean;
(ii) the standard deviation;

(b) explain the meaning of the standard deviation related to the above example.   (CIMA)

26. Consider the following frequency distribution of weights of 150 bolts:

Weight (ounces)	Frequency
5.00 and less than 5.01	4
5.01 and less than 5.02	18
5.02 and less than 5.03	25
5.03 and less than 5.04	36
5.04 and less than 5.05	30
5.05 and less than 5.06	22
5.06 and less than 5.07	11
5.07 and less than 5.08	3
5.08 and less than 5.09	1

(a) Calculate the arithmetic mean and standard deviation of the weights of bolts to three decimal places.
(b) Estimate the number of bolts which are within one standard deviation of the mean.
(c) Suppose that each bolt has a nut attached to it to make a nut-and-bolt. Nuts have a distribution of weights with a mean of 2.043

ounces and standard deviation 0.008. Calculate the standard deviation of the weights of nuts-and-bolts. *(ACA)*

27. (*a*) What is meant by the term 'skewness'?

(*b*) Demonstrate graphically the relationship between the mean, median and mode in both positively and negatively skewed distributions.

(*c*) (*i*) Using the Pearsonian measure of skewness, calculate the coefficient for a frequency distribution with the following values:

mean	15
median	16
standard deviation	6

(*ii*) What conclusion do you reach about this frequency distribution?

(*iii*) Using your own figures or adding to the figures given in (*c*) (*i*), calculate another measure of skewness.

(*d*) Give a practical example of a situation in which identification of a measure of skewness of the distribution would be valuable. *(CIMA)*

28. (*a*) Use the method of least squares to determine the equation of the straight line which best fits the following data. The independent variable is *x*.

x	11	13	14	17	18	21	26
y	20	23	25	28	30	34	38

All calculations must be shown.

(*b*) Draw a scatter diagram for the data and plot the least squares regression line on the scatter diagram. *(CIMA)*

29. (*a*) The least squares line of regression of *y* upon *x* is often used in order to predict the value of *y* given the value of *x*. Explain with the aid of a diagram, what is meant by the term 'least squares'.

(*b*) The following data refers to the time taken to perform a particular operation and its cost. Determine the least squares regression line of cost upon time taken, and illustrate graphically how well your line fits the data. Indicate on your graph the maximum error that would have been made if the regression line had been used to estimate the cost of these operations.

time (mins)	cost (£)
4	7
4	4
5	7
6	9
7	10
8	14
10	18
12	18

(ACA)

30. The following information has been extracted by a marketing organisation:

Weekly household spending by television advertising area

Area	Average total expenditure (£)	Average expenditure on leisure pursuits (£)
London	81	15
Midlands	75	15
Lancashire	73	15
Yorkshire	74	15
North East	74	16
C. Scotland: Grampian	80	17
Wales and West	69	14
South	78	15
Anglia	79	18
South West	72	14
Border	78	17
Ulster	79	18

You are required:

(a) using the least-squares method and applying it to the data above, to estimate the national weekly expenditure per household on leisure pursuits in three years' time when it is estimated the average household total expenditure will be £100 per week (it may be assumed for the purpose of this section that there are no regional differences in the relationship between average total expenditure and average expenditure on leisure pursuits);

(*b*) to comment on the statement that 'people in Wales and West spend far less on leisure than people in Anglia';

(*c*) to distinguish between the terms 'regression' and 'correlation'.

(*CIMA*)

31. (*a*) What is meant by (*i*) correlation, (*ii*) regression?

(*b*) When would it be more appropriate to use regression rather than correlation to measure the degree of association between two variables?

(*c*) Calculate from the following data relating to the total weekly expenditure (x) of ten households in 1978 and their expenditure on housing (y) *either* the coefficient of correlation *or* the equation of the regression of y upon x:

x (£)	98	78	74	80	80	83	95	100	97	75
y (£)	14	9	10	11	10	11	12	13	11	9

(*d*) Comment upon your result in (*c*). (*CA*)

32.

	Index of national average earnings	Annual sales of company (£m)
19–0	80	80
19–1	90	80
19–2	100	90
19–3	110	100
19–4	120	110
19–5	130	120
19–6	150	140
19–7	180	160
19–8	230	190
19–9	260	240

You are required to:

(*a*) calculate a statistic which will provide some indication of the association or relationship between these two series; and

(*b*) discuss the usefulness of such a statistic and consider the extent to which, if at all, it may be used to estimate the relationship between future turnover of your company and the rate of growth in average earnings in the country.   (CA)

33. The annual turnover and net profits of a company over the past decade are as follows:

Year	Turnover (£m)	Profit (£m)
1	60	3
2	75	5
3	90	7
4	120	10
5	140	13
6	145	16
7	150	12
8	160	15
9	170	18
10	140	20

You are required to:

(a) calculate the regression equation of profit on turnover;

(b) estimate from the equation the profit expected in the next two years if turnover rises by 10 per cent in each year; and

(c) comment on the reliability of your results in (b).   (CA)

34. A company has found that the *trend* in the quarterly sales of its furniture is well described by the regression equation

$$y = 150 + 10x$$

where y equals quarterly sales (£000)

        $x = 1$ represents the first quarter of 1980

        $x = 2$ represents the second quarter of 1980

        .

        .

        $x = 5$ represents the first quarter of 1981, etc.

It has also been found that, based on a multiplicative model, i.e.

    Sales = Trend × Seasonal × Random

the mean seasonal quarterly index for its furniture sales is as follows:

Quarter	1	2	3	4
Seasonal index	80	110	140	70

You are required:

(*a*) to explain the meaning of *this* regression equation, and *this* set of seasonal index numbers;

(*b*) using the regression equation, to estimate the trend values in the company's furniture sales for each quarter of 1985;

(*c*) using the seasonal index, to prepare sales forecasts for the company's quarterly furniture sales in 1985;

(*d*) to state what factors might cause your sales forecasts to be in error. (*CIMA*)

35. A department in your factory manufactures product A, which was launched successfully five years ago. The following table gives details of sales in units and total costs since the product was launched:

Year	Sales in units	Total costs £
1	6,000	12,100
2	8,000	15,290
3	10,500	20,500
4	12,500	26,040
5	15,000	31,500

It is now necessary to forecast sales and profit for the next three years because of the possible need for further manufacturing facilities. The total costs above have been affected by inflation as follows:

Year	Index
1	100
2	110
3	125
4	140
5	150

You are required to fit the least squares (regression line) to the cost data (adjusted to year 5 level) and estimate the total fixed costs and the variable cost per unit at year 5 cost levels. (*CIMA*)

36. The time taken to produce batches of an item is found to be related to the number of items in the batch. Records show the

production times for 6 batches to be as follows:

Batch size	Production time (hours)
50	4.0
90	5.8
150	6.8
220	7.6
280	7.6
350	8.6

Two alternatives are suggested for estimating the production time, t, from the batch size, r.

(i) $t = 3.5 + r/50$

(ii) $t = 2 + \sqrt{(r/8)}$

(a) For each of the two relationships find the values of t corresponding to $r = 72$, 128, 200 and 288.

(b) Use the values obtained in (a) to construct a graph illustrating the two alternative relationships. From observing your graph state which best represents the data given in the table.

(c) Find the best fitting curve of the form $t = a + b\sqrt{r}$ using least squares regression of t on $\sqrt{r}$. (ACA)

37. A national consumer protection society investigated seven brands of paint to determine their quality relative to price. The society's conclusions were ranked as follows:

Brand	Price per litre £	Quality ranking
T	1.92	2
U	1.58	6
V	1.35	7
W	1.60	4
X	2.05	3
Y	1.39	5
Z	1.77	1

Using Spearman's rank correlation coefficient, determine whether the consumer generally gets value for money. (CIMA)

38. Ten petrol stations, A–J, which are situated in areas of similar traffic density are ranked first according to quality of service, second according to the size of the forecourt and third according to the price of the petrol sold. Rank 1 indicates best service, largest forecourt and lowest price. The results, including the average weekly petrol sales, are given below.

Station	Quality of service	Forecourt size	Petrol price	Sales (hundreds of gallons)
A	3	8	2	47
B	7	4	9	20
C	4	10	8	23
D	8	2	1	36
E	2	1	4	36
F	5	3	5	31
G	10	9	7	33
H	9	6	10	28
I	1	4	3	22
J	6	7	6	24

(a) Carry out calculations to decide whether price of petrol or quality of service is likely to be the more important factor in determining the volume of petrol sales, or whether neither appears to be important.

(b) Is there any evidence to suggest that those stations with large forecourts give better service?

(c) For what reason do you think that the above analysis has been confined to stations having similar traffic density?   (ACA)

39. A factory produces togs, clogs and pegs, each of these three products having a different work content. The proportions of these products vary from month to month and the factory requires an index for assessing productivity changes. Each tog, clog and peg produced is to be weighted according to its work content, these weights being 6, 8 and 5 respectively. Also, because some months contain more working days than others the index should offset the effect of this.

Data for the months of May, June and July is as follows:

	May	June	July
Number of working days	23	22	16*
Output (thousands):			
Togs	19	16	10
Clogs	12	20	15
Pegs	22	15	10

* Due to factory closure for 2 weeks.

It is intended that May should be the base month for comparison, with a productivity index of 100.

(a) Design a simple productivity index, calculate its value for June and July, and comment briefly on the results.

(b) Now, due to a change in the type of peg produced, a new weight is required. Production data is shown below for two days when productivity was judged to be about equal.

	Day 1	Day 2
Output:		
Togs	921	811
Clogs	800	747
New pegs	1,042	1,206

Use this data to estimate a suitable weight for the new pegs, to 1 decimal place, assuming that the weightings of 6 for togs and 8 for clogs are as before.   (ACA)

40. (a) A company manufacturing a product known as K257 uses five components in its assembly.

The quantities and prices of the components used to produce a unit of K257 in 1982, 1983 and 1984 are tabulated as follows:

	Production of 1 unit of K257					
	1982		1983		1984	
Component	Quantity	Price £	Quantity	Price £	Quantity	Price £
A	10	3.12	12	3.17	14	3.20
B	6	11.49	7	11.58	5	11.67
C	5	1.40	8	1.35	9	1.31
D	9	2.15	9	2.14	10	2.63
E	50	0.32	53	0.32	57	0.32

Required

(*i*) Calculate Laspeyre type price-index numbers for the cost of 1 unit of K257 for 1983 and 1984 based on 1982.

(*ii*) Calculate Paasche type price-index numbers for the cost of 1 unit of K257 for 1983 and 1984 based on 1982.

(*iii*) Compare and contrast the Laspeyre and Paasche price-index numbers you have obtained in (*i*) and (*ii*).

(*b*) A number of employers manufacturing plastic components used in plumbing have formed themselves into an association for the purpose of negotiating with the trade union for this industrial sector.

The negotiations cover pay and conditions in this sector.

Required

Explain the usefulness of an index of industrial production and an index of retail prices to both sides in a series of pay negotiations.

(*ACA*)

41. A survey of household expenditure shows the following changes over the same week in each of three years for an average family in the south of England.

	19–3			19–4			19–5		
	Price in pence	*Unit*	*Quantity purchased*	*Price in pence*	*Unit*	*Quantity purchased*	*Price in pence*	*Unit*	*Quantity purchased*
Sugar	10	2 lb	4 lb	11	2 lb	4 lb	29	2 lb	3 lb
Bread	11	loaf	4	12	loaf	4	16	loaf	4
Tea	8	$\frac{1}{4}$ lb	$\frac{1}{2}$ lb	9	$\frac{1}{4}$ lb	$\frac{1}{2}$ lb	10	$\frac{1}{4}$ lb	1 lb
Milk	5	pint	20	5	pint	21	$5\frac{1}{2}$	pint	19
Butter	9	$\frac{1}{2}$ lb	$1\frac{1}{2}$ lb	10	$\frac{1}{2}$ lb	$1\frac{1}{2}$ lb	13	$\frac{1}{2}$ lb	1 lb

(*a*) Using 19–3 as base, calculate index numbers for 19–4 and 19–5 using the Laspeyre base weighted formula.

(*b*) State, with reasons, whether a survey based on these items represents a reasonable assessment of changes in the cost of living over the three years. (*CIMA*)

42. U Limited, a member of H Limited's group of companies, has been instructed to value its plant and machinery at replacement values

for balance sheet purposes as at 31 December, year 10. H Limited has produced a series of index numbers (average for year 5 = 100) to be used for the purpose of revaluing all plant and machinery.

U Limited has prepared a schedule of its plant and machinery as given below. The index numbers compiled by H Limited for replacement valuations are also shown.

Year of purchase	Original cost £	Aggregate depreciation £	Index number, average for year
1	3,280	2,950	75
2	4,760	3,800	83
3	8,970	6,280	90
4	3,760	2,250	96
5	7,000	3,500	100
6	6,170	2,470	110
7	9,980	2,860	134
8	12,720	2,520	151
9	15,850	1,990	162
10	20,630	1,130	170
	£93,120	£29,750	

The index number at 31 December, year 10 = 175.

(a) Assuming there were no disposals, calculate the cost of replacing U Limited's plant and machinery at 31 December, year 10, based on the information and index numbers provided.

(b) Calculate the aggregate depreciation that would need to be written off if depreciation is based on cost of replacement as in (a) above.

(c) Comment on the revaluation exercise, giving your opinion on the results.   (CIMA)

43. L purchased in the mid-1960s a declining business trading in rubber.

In 1968 he sold his rubber business and bought an oil refinery which traded successfully until the oil crisis of 1973. At about this time he decided to diversify and invested in textiles, and later in the

production of small boats. In 1975 he purchased an electronics
business.

In 1978 L appointed M to be the management accountant of the
group. M calculated and presented at an operations meeting annual
index numbers, using the Paasche method, to show the growth of
volume of the business as a whole since 1975, based on the following
information:

Year	Oil		Textiles		Boats		Electronics	
	units	value	units	value	units	value	units	value
1970	2.0	1.663	–	–	–	–	–	–
1971	2.2	1.888	–	–	–	–	–	–
1972	2.3	2.340	0.8	0.075	–	–	–	–
1973	2.5	2.542	2.7	0.263	–	–	–	–
1974	2.1	4.270	3.5	0.341	0.8	0.111	–	–
1975	2.0	4.060	5.8	0.522	1.6	0.304	10.5	2.730
1976	2.0	4.101	6.9	0.682	2.1	0.482	22.0	5.667
1977	1.7	3.384	10.2	1.108	2.2	0.417	37.6	6.989
1978	1.3	2.388	15.0	1.717	1.9	0.341	59.2	11.004
1979	1.3	2.382	16.8	1.923	1.9	0.310	61.7	11.470

All values in millions of $, and volume in appropriate units in each
industry.

(*a*) Calculate for the total business, the Paasche index number for
1979 as produced by M (1975: 100).

(*b*) Explain briefly how you would have replied to the following
questions asked at an operations meeting:

(*i*) by the operations manager: 'We reorganised this business in
1968. Why did you not use 1968 or 1970 as your base year?'

(*ii*) by the managing director: 'I remember learning to use
Laspeyre's method when I was young and it always proved to be
adequate. Why was the Paasche method used?'

(*iii*) by the sales manager: 'The original information showed
production for 1975 as £7.6 million and for 1979 as £16.1 million.
This I make nearly a 2.2-fold increase. Why calculate an index
number which shows nearly a threefold increase and, I believe,
confuse us?'

Note: Paasche index numbers are: 1976, 141; 1977, 191; and 1978, 259.   (*CIMA*)

44. (*a*) Outline *briefly* the main problems which arise in the construction of an index number. Illustrate your answer by reference to any official index number.

(*b*) The table below shows the prices of each of the five products sold by a company and the proportionate contribution the sales of each make to total turnover in 19–4 and 19–9.

Calculate an index number to measure the overall change in the prices of these products between 19–4 and 19–9. Do you consider that the choice of the base year for the index poses any problems? If so, how do you suggest they might be dealt with given that the company wishes to continue with this index in future years?

			Product		
	A	B	C	D	E
19–4:					
Price £	1	2	4	6	10
% of sales	5	10	20	30	35
19–9:					
Price £	2	4	6	10	25
% of sales	15	20	20	25	20

(*CA*)

45. In the introduction to the publication *Price Index Numbers for Current Cost Accounting* it is stated that each price (of each item of plant, machinery etc.) is expressed as a percentage of the corresponding price in a fixed base period. The resultant price relatives are then combined by means of a weighted arithmetic average where the weights are, so far as possible, proportional to the composition, by value, of plant and machinery bought by the specified industry. The resulting figure is then multiplied by a constant scaling factor to make the index equal to 100 in the base period.

An important industrial client has asked for a 'simple' explanation of what the above passage means and what purpose it is supposed to serve. Draft, in the form of a letter, an appropriate explanation.

(*CA*)

46. The Sandilands Report on inflation accounting contains the following statement:

'Just as a general index of prices is of little practical use to any particular individual or entity so a general index of the "purchasing power" of money is unlikely to be helpful.'

You are required to:

(a) explain and discuss this statement; and
(b) suggest, with reasons, a method for adjusting accounts for inflation in the light of the difficulties of using a general index.

(CA)

47. Quarterly indices of retail sales in Great Britain 19–1 to 19–4.

| Year | Quarters | | | |
	Winter	Spring	Summer	Autumn
19–1	91	94	98	118
19–2	99	105	110	134
19–3	114	118	124	152
19–4	122	130	140	161

You are required to:

(a) calculate the quarterly movement of this series;
(b) derive the trend;
(c) graph the data and insert the trend line on the graph; and
(d) comment upon the result.

(CA)

48. The following information has been supplied by the sales department.

Sales in units

| Year | Quarter | | | |
	1	2	3	4
19–3	100	125	127	102
19–4	104	128	130	107
19–5	110	131	133	107
19–6	109	132		

You are required to:

(a) calculate a four-quarterly moving average of the above series;

(b) calculate the sales corrected for seasonal movements;

(c) plot the actual sales and the sales corrected for seasonal movements on a single graph; and

(d) comment on your findings. (*CIMA*)

49. From the data supplied calculate the trend, the seasonal or, in this case, average daily variations, and the residuals. The figures represent cash sales of a store which does not open for business on a Monday.

	Week number			
	1	*2*	*3*	*4*
Day of week	£	£	£	£
Tuesday	360	350	380	390
Wednesday	400	430	440	450
Thursday	480	490	490	500
Friday	600	580	590	600
Saturday	660	680	690	690

(*CIMA*)

50. The sales of an item which show a marked seasonal variation are given below for the last three years.

	Sales (*thousands*)					
Year	Jan/Feb	Mar/Apr	May/Jun	Jul/Aug	Sept/Oct	Nov/Dec
19–3			10.3	12.8	20.3	32.4
19–4	6.8	9.0	10.9	14.4	23.1	35.4
19–5	7.0	9.5	11.6	15.3	24.5	37.4
19–6	7.3	9.9				

(a) Use the method of moving averages to find the trend.

(b) Assuming that the general rate of increase of sales occurring over the period observed is maintained, estimate graphically the trend values for each of the remaining two-month periods of 19–6.

(c) Determine the seasonal effect for the period July/August and estimate the sales that will occur in the period July/August 19–6.

(*ACA*)

51. A medium-sized manufacturing organisation has the following sales analysis for the period 1971–80 inclusive.

Year	Sales £ million
1971	15.3
1972	14.6
1973	16.8
1974	17.3
1975	17.2
1976	20.9
1977	22.3
1978	20.0
1979	23.1
1980	24.5

You are required to:

(a) calculate:

(i) the trend line using the least squares equation;

(ii) the trend values for 1971 and 1980 and explain the meaning of these results;

(b) draw a graph of the above data including the trend and project the trend for the year 1981. (CIMA)

52. The annual sales of Steady Growth Ltd in each of the ten years 1966–75 have been as follows:

(£000s)   5,   9,   13,   17,   25,   30,   34,   40,   46,   54

You are required to:

(a) estimate the annual sales in 1978 and in 1981 on the assumption that the company will maintain the same rate of growth as in the past decade; and

(b) comment on the validity of your estimate in the light of this series and of the assumption. (CA)

53. You are required to calculate all of the following probabilities.

(a) Calculate the probability of getting, with two throws of a die:

(*i*) a score of less than 9; and

(*ii*) a total score of 9.

(*b*) Four different coloured balls, red, blue, white and yellow, are contained in a bag. The balls are drawn singly at random. Calculate the probability that:

(*i*) the first ball will be blue; and

(*ii*) the last ball will be red.

(*c*) Tickets were sold in a raffle from four different coloured books in these numbers: 124 red; 136 green; 96 blue; and 85 white. Calculate the probability that:

(*i*) a person who had bought five tickets wins the raffle; and

(*ii*) the winning ticket is white.

(*d*) In an examination where 150 candidates took two papers, economics and statistics, the failure rate on each was 30 and 25 per cent respectively. Calculate the probability that a candidate chosen at random would have:

(*i*) failed both papers;

(*ii*) passed both papers; and

(*iii*) failed economics but passed statistics.

(*e*) There are known to be two defective fuses in a box of ten fuses. Calculate the probability of finding the two defective fuses in testing the first five.

(*f*) The chances of a man aged 55 dying within the next five years are 0.02 and a woman of the same age are 0.01. Calculate the probability that a husband and wife who reach their 55th birthday on the same date will by their 60th birthday:

(*i*) both be alive;

(*ii*) both be dead; and

(*iii*) one of them will be alive and the other dead.   (*CA*)

54. (*a*) An insurance agent finds that on following up an inquiry the probability of making a sale is 0.4. If on a particular day the agent has two independent inquiries what is the probability that he will sell:

(*i*) insurance to both inquirers;

(*ii*) exactly one policy;

(*iii*) at least one policy?

(*b*) An oil company drilling in the North Sea reckoned that there was a 60 per cent chance of success of finding oil in economic quantities in a particular field. After a first test drilling had been completed the results were favourable.

If the probability is 0.3 that the test drilling would give a misleading result, what would be the revised probability, using Bayes' Theorem, that a find of oil in economic quantities would occur?   (*CIMA*)

55. (*a*) Calculate the probability of drawing three picture cards of a black suit from a pack of 52 playing cards.

(*b*) Calculate the probability that when two dice are rolled the sum of the spots showing will be (*i*) even, (*ii*) less than 8, (*iii*) exactly 11 and (*iv*) exactly 12.

(*c*) Five per cent of the population are allergic to a particular substance. Tests to determine whether a person is allergic are only 90 per cent reliable. What is the probability that a person tested for this allergy by this technique will show a positive reaction?

(*d*) Ten per cent of a large bag of marbles are black. What is the probability if four marbles are selected at random that (*i*) exactly 2 are black, (*ii*) that at least 1 is black?

(*e*) Assuming the wages of typists are normally distributed, with a mean of £80 and a standard deviation of £8, what proportion of typists will earn between £90 and £95?   (*CA*)

56. (*a*) What is the probability of throwing at least one 6 in two throws of a die?

(*b*) The successful operation of three separate switches is needed to control a machine. If the probability of failure of each switch is 0.1, what is the probability that the machine may break down?

(*c*) Calculate the probability that, when two dice are rolled, the sum of the spots showing will be:

(*i*) an odd number;
(*ii*) less than 9;
(*iii*) exactly 12;
(*iv*) exactly 4.

(*d*) Twenty per cent of a large bag of marbles are black in colour. What is the probability if 5 marbles are drawn at random (with replacement) that:

(*i*) all 5 are black;

(*ii*) at least 3 are black.

(*e*) The mean height of a sample of 18-year-old males is 68″ and the standard deviation of the distribution is 1″. What proportion of the sample will be,

(*i*) between 65″ and 68″ tall;

(*ii*) over 65″ tall;

(*iii*) over 70″ tall? (*CA*)

57. (*a*) Three machines, A, B and C, produce 60 per cent, 30 per cent and 10 per cent respectively, of the total production. The percentages of defective production of the machines are 2 per cent, 4 per cent and 6 per cent respectively.

If an item is selected at random, find the probability that the item is defective.

(*b*) Using the information given in part (*a*) and assuming an item selected at random is found to be defective, find the probability that the item was produced on machine A. (*CIMA*)

58. You are required to calculate the probabilities in each of the following problems.

(*a*) Three components A, B and C are produced by a manufacturer and later assembled. Experience has shown that the proportion of defectives in the output of the three components is 1, 2 and 5 per cent, respectively. What are the chances that upon assembly:

(*i*) components A and B are both faulty; and

(*ii*) any one of the three components is faulty?

(*b*) A bag contains 100 coloured balls; 25 red, 25 yellow and 50 blue. What is the probability that:

(*i*) the first ball drawn from the bag will be red or yellow;

(*ii*) the first ball is red and the second blue; and

(*iii*) that the first three balls will be of different colours?

(*c*) (*i*) A die is thrown 24 times and yields only one 6. What is the probability that on the 25th throw a 6 will show?

(*ii*) Whatever the result obtained, is there reason for believing that the die is not properly balanced?

(*d*) A company instals vending machines and finds that, on average, 40 per cent of the installations need to be further adjusted after installation. If four machines are installed in a customer's premises what is the probability that at least two will need further adjustment?

(*CA*)

59. (*a*) You are required, for two events *a* and *b*, to express $P(a \cup b)$ in terms of $P(a)$ and $P(b)$ only, when:

(*i*) the events *a* and *b* are mutually exclusive;
(*ii*) the events *a* and *b* are independent.

Note: $P(a \cup b)$ is also written as $P(a$ or $b)$.

(*b*) A manufacturer has three main suppliers of fuses. The table below shows the quantities of colour-coded fuses held in stock by the manufacturer. As part of a quality control check a sample of boxes of fuses is to be tested. Boxes of fuses are of identical size, each box containing 100 fuses.

Number of boxes of fuses in stock

Colour code	Red	White	Green	Blue	Totals
Supplier					
X	100	90	110	100	400
Y	40	60	50	50	200
Z	10	50	190	150	400
Totals	150	200	350	300	1,000

You are required:

(*i*) If a sample of 100 boxes is to be checked for quality, to state how you would recommend the sample to be selected. (You may assume that fuses cost equal amounts to check.)

(*ii*) If *one* box is to be selected randomly from the total, to state the probability of it being:

(1) red;
(2) from Supplier X;
(3) red and from Supplier X;
(4) red or from Supplier X.

(*iii*) If *two* boxes are to be chosen randomly from the total, to state the probability that they will both be from the same supplier.

(*iv*) One box having been selected randomly from the total and given that it has been identified as having come from Supplier Y, to state the probability it contains fuses that are *not* red.　(*CIMA*)

60. (*a*) What is the binomial distribution? What is its relevance to statistical sampling theory?

(*b*) A die is thrown four times and either a 5 or 6 is regarded as a success. Calculate the probability distribution of successes.

(*c*) What is the probability of obtaining more than two sixes in four throws of a die?　(*CA*)

61. (*a*) A fair die with six sides is thrown three times. Show by means of a tree diagram that the probability of obtaining 0, 1, 2 or 3 sixes from the three throws is given by the binomial probability function:

$$\binom{3}{r}\left(\frac{1}{6}\right)^r\left(\frac{5}{6}\right)^{3-r}$$

where *r* represents the number of successes.

(*b*) A department produces a standard product. It is known that 60 per cent of defective products can be satisfactorily reworked.

What is the probability that in a batch of five such defective products at least four can be satisfactorily reworked?　(*CIMA*)

62. (*a*) A company has invested £50,000 in a plant which should meet all foreseeable demand for a patented product. The annual demand is known to be approximately Normally distributed with a mean of 20,000 units and a standard deviation of 4,000 units. If the contribution on each unit sold is £2, what is the probability that the company will recover its investment within one year?

(*b*) An Examining Board has to compare the *overall* performance of three candidates A, B and C who sat four subjects of equal importance. Their marks are given below, together with the mean and standard deviation of the results of all (one thousand) candidates.

	Candidate			All Candidates	
Subject	A	B	C	Mean	SD
Accounting	60	55	65	45	10
Economics	47	43	51	35	8
Law	75	83	70	60	15
Statistics	70	80	65	50	20

Assuming each subject to have an approximately Normal distribution of marks, find which candidate(s) has the best *overall* performance.   (*CIMA*)

63. A large batch of items comprises some manufactured by process x and some by process y. There are twice as many items from x as from y in the batch. Those from x contain 9 per cent defectives and those from y contain 12 per cent defectives.

(*a*) Calculate the proportion of defective items in the batch.

(*b*) Calculate the probability that 3 items taken at random from the batch contain 1 defective.

(*c*) An item is taken at random from the batch and found to be defective. What is the probability that it came from y?

(*d*) It is found that 50 per cent of defective items cannot be repaired and must be replaced. The cost of repairing a defective is £2 if it comes from process x and £3 if it comes from y. Replacements cost £7 each. Calculate the expected cost of repair and replacement of the defectives found in 300 items taken from the batch.   (*ACA*)

64. In an acceptance sampling scheme, a random sample of size 100 items is taken from a batch and inspected, and if it contains 3 or fewer defectives the batch is accepted without further inspection, and is said to have passed the sampling scheme. If the sample contains 4 or more defectives then all the remaining items in the batch are inspected and the batch is said to have failed the scheme.

(*a*) By using the Poisson distribution to calculate the probability of finding 0, 1, 2 and 3 defectives, determine the proportion of batches containing 2 per cent defective items that will pass the acceptance sampling scheme.

(*b*) Show that the expected number of defectives found by the scheme in a batch containing 2 per cent defectives that passes the scheme is approximately 1.6. By considering, in addition, those batches that fail the scheme, determine the expected number of defectives found in batches of size 2,000 containing 2 per cent defectives. (*ACA*)

65. The life of a television (TV) tube, measured in hours of use, is approximately Normally distributed with a mean of 5,000 hours and a standard deviation of 1,000 hours.

(*a*) Calculate the probability that a TV tube will last for

  (*i*) between 5,500 and 6,500 hours;
  (*ii*) less than the expected life of a tube.

(*b*) Calculate the probability that, for someone using a TV for 1,500 hours per year, the tube will last for less than five years.

(*c*) Suppose that the company offers a guarantee to customers, at an additional charge of £25. Under the guarantee, the TV tube will be replaced free of charge if it fails within three years of purchase and will be replaced at a cost of £40 to the customer if it fails between the end of the third year and the end of the fifth year.

The cost of a TV tube to the company is £80.

Assuming that a TV is used for 1,500 hours per year, calculate the expected cost to the company through offering a guarantee on a tube. (*ACA*)

66. (*a*) What is the Normal distribution? What is its significance for statistical sampling?

(*b*) 28 per cent of a certain population is estimated to have an IQ of 95 or less. A further 12 per cent is estimated to have an IQ of 130 or more. Assuming that the characteristic of intelligence is Normally distributed, what is the mean and standard deviation of this population? (*CA*)

67. The label on a container of a liquid chemical sold at 10 pence per litre states the contents to be 10 litres. However, the filling equipment cannot fill each container with exactly the same volume of liquid: the volumes are Normally distributed with a standard deviation of 0.2 litres. The latter may be treated as a constant, but the mean fill (at present 10 litres) can be adjusted. Regulations require that no more than one in a hundred containers should contain less than 10 litres.

A company has designed a modification which can be fitted to the filling equipment and would reduce the standard deviation to 0.15 litres. It would cost £5,000 for the modification which would need to be replaced after 100,000 containers had been filled.

You are required to:

(*a*) state the level at which the mean fill should be set in order to meet the regulation, without using the modification;

(*b*) advise management if the modification is a worthwhile purchase. (*CIMA*)

68. An assembly produced in large quantities is made up of three different components, A, B and C. The weight of each component is approximately Normally distributed with mean and standard deviation as follows:

Component	Weight (pounds)	
	Mean	Standard deviation
A	3.0	0.2
B	2.5	0.2
C	6.0	0.4

(*a*) What proportion of assemblies

(*i*) exceed 11.8 pounds?
(*ii*) are between 11.4 and 11.7 pounds?

(*b*) Any assembly weighing more than 11.8 pounds is unsatisfactory and is scrapped. Determine the median weight of satisfactory assemblies. (*ACA*)

69. (*a*) The central limit theorem states that, as n increase, the distribution of sample means is normal with mean μ and standard error $\sigma/\sqrt{n}$.

Explain briefly the meaning of the terms:

(*i*) distribution of sample means;
(*ii*) standard error.

(*b*) Items are manufactured to a mean weight of 3 lb and a standard deviation of 0.05 lb. Cartons each containing nine items are sold on the understanding that the mean weight of items in a carton is not less than 2.97 lb. Cartons not meeting this requirement are rejected.

(*i*) Calculate the proportion of cartons which are rejected.

(*ii*) Suppose now that the proportion rejected can be reduced by adjusting the process so that the mean weight of all items is increased to 3.01 lb. This adjustment costs £0.36 per full carton. If the cost of a rejected carton is £10, calculate whether the adjustment is economically desirable. (*ACA*)

70. (*a*) Items produced to a weight specification are rejected if found to be outside the range 2.975 to 3.025g. A batch of 500 items is found to have weights which are Normally distributed with a mean of 3.005 and a standard deviation of 0.015. How many items are rejected?

(*b*) The standard procedure used by a company for detecting defects in a product has, from experience, been shown to detect 90 per cent of defects present.

(*i*) Calculate the standard error of the proportion of defects detected out of 100 examined and find the probability that, out of 100 defects, 15 or more are undetected by the procedure.

(*ii*) An alternative procedure is suggested which during trials detects 368 out of 400 defects. Show that the new procedure does not give a statistically significant improvement in detection rate. (*ACA*)

71. (*a*) What do you understand by the terms:

(*i*) statistically significant; and
(*ii*) confidence interval.

(*b*) A survey among a sample of 200 employees in a northern factory reveals that the mean gross wage is £95 per week with a standard deviation of £15. A survey among all the factories controlled by this company revealed that average earnings were £90 per week with a standard deviation of £10. Test the hypothesis that earnings in the northern factory are significantly higher than over the group. Explain the reasoning underlying your test. (*CA*)

72.

Distribution of weekly household income in the UK 1975

£ per week	% of households
Under 15	6.1
15 and under 20	6.8
20 and under 30	9.8
30 and under 40	8.8
40 and under 50	9.4
50 and under 60	10.2
60 and under 70	10.0
70 and under 80	8.9
80 and under 90	7.3
90 and under 100	5.6
100 and under 120	7.6
120 and over	9.5

Source: Social Trends 1976

You are required to:

(*a*) calculate the mean and standard deviation of the above distribution;

(*b*) estimate the standard error of the mean, given that the data is from a random sample of 7,200 households; and

(*c*) test the hypothesis that the difference between the UK mean income and that of the West Midlands region of £66.70, based upon random sample of 670 households, is not statistically significant.

(*CA*)

73. (*a*) A factory has ten identical drilling machines numbered 7601 to 7610. The summary of the maintenance department records for 19–6 showed the following days lost (*See* page 451):

You are required to state the average number of days that it is expected will be lost in 19–7. Place the 95 per cent confidence limits on this average.

Machine number	Days lost
7601	15
7602	16
7603	18
7604	13
7605	21
7606	14
7607	15
7608	17
7609	23
7610	18

(b) The analyses of sales for two sales managers for 19–6, based on random samples, were:

Annual sales by customer turnover	Percentages of customers	
	Midland	South-West
< £10	1	1
£10 and < £20	10	1
£20 and < £30	12	8
£30 and < £40	10	14
£40 and < £50	6	17
£50 and < £60	8	16
£60 and < £70	8	20
£70 and < £80	4	12
£80 and < £90	5	4
£90 and <£100	6	5
£100 and <£300	30	2
	100	100

Additionally:

$\Sigma fx = 9,330 \quad 5,770$
$\Sigma fx^2 = 1,404,950 \quad 411,250$

You are required to test the hypothesis that there is no difference in the average annual sales per customer between the two areas.

(*CIMA*)

74. (*a*) You are required to:

(*i*) define and explain the term 'confidence interval'; and
(*ii*) calculate the 95 per cent confidence interval where 52 per cent of a sample of 800 consumers state a preference for a given product against its competitors. Interpret your results.

(*b*) In a controlled experiment the success rate of a new drug in curing a sample of 200 patients suffering from a certain disease is 60 per cent. The drug currently in use has a comparable success rate of 54 per cent. Are the doctors justified in inferring that the new drug is significantly superior as a treatment? Explain carefully how you interpret the result of your statistical test. (*CA*)

75. A company selling washing machines finds that its established salesmen achieve average sales of 5.5 machines a week. Any new salesman is given a trial for 5 weeks and is not taken on permanently unless he sells an average of at least 4.8 machines a week. A new salesman sells 5, 3, 4, 6 and 5 and fails the trial. He claims that the system is unfair and that he can perform in the long run as well as the established salesmen, and yet fail the trial badly.

(*a*) Carry out appropriate calculations and comment on his claim.
(*b*) Calculate the minimum average weekly sales that should be required on a 5 weeks' trial if a new salesman is to be rejected only if his performance is significantly different from 5.5. (*ACA*)

76. (*a*) A consumer survey among urban housewives entailed two samples of 800 in a southern and northern city. The proportion buying a particular product in the south was 50 per cent; in the north it was 45 per cent. The marketing manager of the producing company claims that this result justifies a new marketing policy based upon the premise that the south is a much more attractive market for the company.

The directors of the company have asked for your views on the marketing manager's claim. Draft a report in which you explain first, the basis of the relevant statistical test to determine that there is evidence of a significant difference between the two markets and second, whether the marketing manager's claim should be accepted and his proposals adopted.

(b) Suppose the same results had been derived from samples of 500 housewives in each city. Would your conclusions be different? Give reasons for your answer. (CA)

Appendix 7
Assignment data

1. Ages of residents (15 and over) in Upper End and Lower End

Age last birthday (yrs)	No. in Upper End	No. in Lower End
15–19	3	1
20–24	20	10
25–29	80	50
30–34	120	60
35–39	90	50
40–44	80	40
45–49	70	30
50–54	55	25
55–59	40	20
60–64	20	10
65–69	10	5
70 and over	4	2

2. Weekly incomes – Moneytown

Weekly income (£)	No.
Under 80	10
80–120	20
120–150	140
150–200	90
200–250	40
250–350	32
350–500	20
500–700	14
700–1000	4

3. Weights of bundles of paper

Weight (lbs)	Bundles
97–under 98	5
98–under 99	12
99–under 100	20
100–under 101	28
101–under 102	22
102–under 103	16
103–under 104	11
104–under 105	5
105–under 106	2

4. Prizes – Lottie's lottery

Prize (£)	No.
1	1,300
5	500
10	120
25	50
50	20
100	7
500	2
1,000	1
	2,000

5. Monthly marriages registered – Seemoff Registry Office

No. in month	Months
Under 5	4
6–10	12
11–15	20
16–20	60
21–25	30
26–30	50
31–35	90
36–40	120
41–45	40
over 45	14

6. Light life. 200 lights were set up on a board and lit. The board was inspected every 50 hours. Results are as follows:

	Check	Lights still burning
At	800 hrs.	200
	850	190
	900	160
	950	100
	1,000	62
	1,050	38
	1,100	20
	1,150	6
	1,200	0

7. TV viewing. 3,500 people were interviewed to see how many episodes they watched of that great TV serial, 'Loder Oletripe'.

No. of episodes viewed	No. of viewers
Under 5	1,400
5–9	600
10–14	200
15–19	400
20–24	900
	3,500

8. House values in Snobslet

Value (£)	No. of houses
Under 55,000	8
55,000–105,000	50
over 105,000	2
	60

9. Journey lengths – analysis of data collected from family drivers

Miles travelled	Average mileage	Journeys
0–under 5	2	220
5–under 20	8	100
20–under 50	30	60
50–under 100	70	10
100–under 200	120	4

Index

M&E Handbooks

Law

'A' Level Law/B Jones
Basic Law/L B Curzon
Cases in Banking Law/P A Gheerbrant, D Palfreman
Cases in Company Law/M C Oliver
Cases in Contract Law/W T Major
Commercial and Industrial Law/A R Ruff
Company Law/M C Oliver, E Marshall
Constitutional and Administrative Law/I N Stevens
Consumer Law/M J Leder
Conveyancing Law/P H Kenny, C Bevan
Criminal Law/L B Curzon
Equity and Trusts/L B Curzon
Family Law/P J Pace
General Principles of English Law/P W D Redmond, J Price, I N Stevens
Jurisprudence/L B Curzon
Labour Law/M Wright, C J Carr
Land Law/L B Curzon
Landlord and Tenant/J M Male
Law of Banking/D Palfreman
Law of Evidence/L B Curzon
Law of Torts/J G M Tyas
Meetings: Their Law and Practice/L Hall, P Lawton, E Rigby
Mercantile Law/P W D Redmond, R G Lawson
Private International Law/A W Scott
Sale of Goods/W T Major
The Law of Contract/W T Major

Business and Management

Advanced Economics/G L Thirkettle
Advertising/F Jefkins
Applied Economics/E Seddon, J D S Appleton
Basic Economics/G L Thirkettle
Business Administration/L Hall
Business and Financial Management/B K R Watts
Business Organisation/R R Pitfield
Business Mathematics/L W T Stafford
Business Systems/R G Anderson
Business Typewriting/S F Parks
Computer Science/J K Atkin
Data Processing Vol 1: Principles and Practice/R G Anderson
Data Processing Vol 2: Information Systems and Technology/R G Anderson
Economics for 'O' Level/L B Curzon
Elements of Commerce/C O'Connor
Human Resources Management/H T Graham
Industrial Administration/J C Denyer, J Batty
International Marketing/L S Walsh
Management, Planning and Control/R G Anderson
Management – Theory and Principles/T Proctor
Managerial Economics/J R Davies, S Hughes
Marketing/G B Giles
Marketing Overseas/A West
Marketing Research/T Proctor, M A Stone
Microcomputing/R G Anderson
Modern Commercial Knowledge/L W T Stafford
Modern Marketing/F Jefkins
Office Administration/J C Denyer, A L Mugridge
Operational Research/W M Harper, H C Lim
Organisation and Methods/R G Anderson
Production Management/H A Harding
Public Administration/M Barber, R Stacey
Public Relations/F Jefkins
Purchasing/C K Lysons
Sales and Sales Management/P Allen
Statistics/W M Harper
Stores Management/R J Carter

Accounting and Finance

Auditing/L R Howard
Basic Accounting/J O Magee
Basic Book-keeping/J O Magee
Capital Gains Tax/V Di Palma
Company Accounts/J O Magee
Company Secretarial Practice/L Hall, G M Thom
Cost and Management Accounting – Vols 1 & 2/W M Harper
Elements of Banking/D P Whiting
Elements of Finance for Managers/B K R Watts
Elements of Insurance/D S Hansell
Finance of Foreign Trade/D P Whiting
Investment: A Practical Approach/D Kerridge
Practice of Banking/E P Doyle, J E Kelly
Principles of Accounts/E F Castle, N P Owens
Taxation/H Toch

Humanities and Science

Biology Advanced Level/P T Marshall
British Government and Politics/F Randall
Chemistry for 'O' Level/G Usher
Economic Geography/H Robinson
European History 1789–1914/C A Leeds
Introduction to Ecology/J C Emberlin
Land Surveying/R J P Wilson
Modern Economic History/E Seddon
Political Studies/C A Leeds
Sociology 'O' Level/F Randall
Twentieth Century History 1900–45/C A Leeds
World History: 1900 to the Present Day/C A Leeds